Reassembling Rubbish

Reassembling Rubbish

Worlding Electronic Waste

Josh Lepawsky

The MIT Press
Cambridge, Massachusetts
London, England

This book was set in ITC Stone Serif Std by Toppan Best-set Premedia Limited. Printed and bound in the United States of America.

Library of Congress Cataloging-in-Publication Data

Names: Lepawsky, Josh, 1972- author.
Title: Reassembling rubbish : worlding electronic waste / Josh Lepawsky.
Description: Cambridge, MA : The MIT Press, [2018] | Includes bibliographical
    references and index.
Identifiers: LCCN 2017039339| ISBN 9780262037884 (hardcover : alk. paper) |
    ISBN 9780262535335 (pbk. : alk. paper)
Subjects: LCSH: Electronic waste.
Classification: LCC TD799.85 .L47 2018 | DDC 628.4/4--dc23 LC record available at
https://lccn.loc.gov/2017039339

10 9 8 7 6 5 4 3 2 1

For those to whom I am kith and kin, especially Mom, Dad, Noah, Erin, and Arlo

Contents

Acknowledgments

I am grateful to the Social Sciences and Humanities Research Council (SSHRC Grant 435–2012–0673) and to Memorial University of Newfoundland for their support, which enabled me to write this book and to research the broader project of which it is part. It would have been impossible to conduct the substantial fieldwork required to follow discarded electronics, support graduate students, or have the necessary time for research and writing without the financial and material support both institutions provided.

Deciding to write this book came not out of any grand plan of my own but as a result of an unexpected invitation from Joshua A. Bell, curator of globalization at the Smithsonian Institution's National Museum of Natural History in Washington, D.C. In March 2014 he asked me to participate in a public event on sustainable design and repair. Crafting the public presentation I would give at that event became the first step in formulating the ideas for this book. My thanks to Joshua for setting off such an auspicious chain of events.

Developing and writing this book project has been a genuinely joyful experience. That writing would become such a pleasure is in no small part due to the fantastic mentorship of people who have taught me so well. Anne Godlewska honed my skills in constructing an argument. She also offered the sagest advice I have ever received from an academic about being an academic: be sure to have fun unrelated to scholarship or academia will eat you alive.

Anne's advice helped me survive the rigors of my PhD training at the University of Kentucky. The program there was intensely challenging, but reflecting on the twelve years that have passed since I completed my doctorate, I would not give up one moment of the time I spent in Lexington.

Sue Roberts was an outstanding supervisor to me there. Her dedication to facilitating my success in the program was of an immense help—especially in my darker moments while suffering from imposter syndrome. John Pickles taught me what it is to read ethically. It was also from him that I began to learn the importance of understanding science and technology's histories and geographies.

Much of the research on which this book is based was carried out in conjunction with graduate students. Chris McNabb and I worked together early on to develop cartograms of e-waste flows. His diligence and technical prowess helped me become familiar with a quantitative world I had no deep background in before we began working together. Mostaem Billah introduced himself to me in an e-mail from Dhaka, Bangladesh, long before I had the resources to support graduate students. He and I would go on to do extensive fieldwork together in the city he knows so well. Doing so on my own would have been impossible. Creighton Connolly's research in Singapore helped me stay connected to developments I had begun tracing there earlier. Grace Akese accompanied my initial attempt at gaining on-the-ground insight into the dynamics of Agbogbloshie in Accra, Ghana. It quickly became apparent that my presence there was more hindrance than help. Yet thanks to her persistence with her own fieldwork, she has helped me situate Agbogbloshie in the worldings of electronic waste others have made for it. It has also been a pleasure to work with Erin Araujo in Mexico and in Peru. Her skills in collecting field data have been instrumental in our joint work as well as in her own efforts. John-Michael Davis has also been key to research conducted in Mexico. Again, doing that research on my own, without either Erin's or John-Michael's assistance, would not have been possible.

Working with colleagues and collaborators on the Reassembling Rubbish project has been essential to gleaning insights from the research for it. Robin Ingenthron offered me portals into the business of the scrap sector and the repair sector that I would never have been able to find, let alone understand, without his help. I appreciate his being a tough audience, and I hope I continue to earn the deep trust he has placed in me. Ramzy Kahhat, an engineer, and Josh Goldstein, a historian of contemporary China, continue to be fantastic scholars to collaborate with. I am deeply thankful for their willingness to engage with me in a highly interdisciplinary project that spills across borders both earthly and conceptual.

I am exceptionally lucky to work with a dedicated group of colleagues on a day-to-day basis at Memorial University of Newfoundland. Marc Bolli at the Core Research Equipment and Instrument Training (CREAIT) Network and Craig Squires at the Atlantic Computational Excellence Network (ACENET) have provided invaluable technical assistance. In the Department of Geography I am particularly grateful to Mario Blaser, Arn Keeling, Max Liboiron, and Charlie Mather, all of whom read early versions of the bits and pieces of manuscript that would become this book. Carissa Brown stepped me through some surprisingly tricky last-minute arithmetic with humor and grace. As the citations in this book attest, I am especially thankful for Max's thinking. It is an amazing gift to be able to knock at the office two doors down and talk about everything I am slow to understand as I pursue my own discard studies.

Erin and Arlo: I love you. Thank you for reciprocating and for being interested in entirely other projects of your own making.

Note on Figures

Many of the figures created for this book invite reader interaction on the book's companion website. The images designed for online interactions do not appear as figures within the book text but are called out in the text, for example, "figure 1.1 online." All figures, organized by chapter, are available on the book's companion website, worldingelectronicwaste.xyz.

1 Introduction

More than fifteen years ago the freelance journalist Russ Arensman wrote a piece for *Electronic Business*, a now defunct trade magazine, about something called "electronic waste" (Arensman 2000, 110). Largely a profile of Hewlett-Packard's (HP's) then new recycling plant in Roseville, California, the article revealed a geography of electronics recycling in which plastics and cathode ray tubes (CRTs) recovered at the Roseville plant were trucked more than 5,000 kilometers northeast to Noranda Inc.'s smelters in Canada (see figure 1.1 online). At the time, Noranda operated the Horne smelter in the company's hometown of Rouyn-Noranda, Quebec. Noranda no longer exists (it was absorbed in a series of mergers with other mining companies, first Falconbridge and later Xstrata), but at the time it also had controlling interests in operations outside Quebec. One of those operations was the Brunswick Mining and Smelting facility near Belledune, New Brunswick, on the shores of Chaleur Bay. CRT monitors were crushed and processed for their copper and lead at both the Horne and Brunswick smelters. The plastics from those monitors and other electronics processed from HP's Roseville plant were burned for fuel at both sites. The year Arensman's article appeared, the Horne smelter emitted 90,000 metric tons of sulfur dioxide, 620 metric tons of particulates, 80 metric tons of lead, and 2.2 metric tons of cadmium. The influence of cadmium emissions was traced as far as 25 kilometers from the smelter; for lead it was up to 300 kilometers (Bonham-Carter 2005, 10; see also Savard, Bonham-Carter, and Banic 2006, 101, 104). The Brunswick smelter's atmospheric emissions for the same year included 11,938 metric tons of sulfur dioxide, 86.01 metric tons of particulates, 13.89 metric tons of lead, 2.43 metric tons of zinc, 1.36 metric tons of cadmium, and 1.49 metric tons of arsenic (Bonham-Carter 2005, 12; see also Parsons and Cranston 2006, 261). Insofar as the Horne smelter had a total throughput of over 787,000 metric tons of feedstock from all sources in the year 2000 (Savard, Bonham-Carter, and Banic 2006, 101), the contribution of the electronics arriving from HP's Roseville recycling plant would have been minuscule (feedstock data are unavailable for the Brunswick smelter). However, the inputs of recycled electronics arriving from California (and elsewhere) into the smelter operations in Canada highlight how recycling electronics is action that connects people, places, and things. The action generating those connections leaves a wake of discards, such as emissions from long-distance truck transport and smelting. Such is the case with practices of discarding such as recycling. Though one might intuitively think of discarding as a ridding that separates (e.g., a person from her possessions), such practices are also groupings—or attachment sites, as Donna Haraway (2008, 2010) might say. In other words, discarding generates partial and specific connections through action carried out, and that action, though partial and situated, raises fundamental ethical questions—questions about good and right action—which are explored in this book.

The transport of discarded electronics between HP's Roseville facility and Noranda's smelters in Quebec and New Brunswick stitched together the waste management sector and primary industry. It was hardly the first time such connections were made. Three decades earlier, as the United States was grappling with the implications of the Solid Waste Disposal Act of 1965 and growing concerns about resource scarcity, researchers at the U.S. Bureau of Mines turned their attention to a largely untapped source of recoverable precious metals: discarded electronics (Dannenberg, Maurice, and Potter 1972; see also Kenahan et al. 1973; Kleespies, Bennetts, and Henrie 1970). In a 1972 report, Dannenberg and co-workers (1972, 1) noted that electronic scrap is "generated in large quantities by military and civilian electronic operations." In their experiments these authors were able to recover silver, gold, and copper at rates of 93, 95, and 87 percent, respectively. But what is important to notice, beyond these impressively efficient recovery rates, is the contrast between how discarded electronics were being enacted (that is, how they were being brought into being through practices; see Mol 2002, 32) by the U.S. Bureau of Mines in the 1970s—as a source of recoverable resources—and how they were enacted in Arensman's article, written some thirty years later, as a problem of post-consumption waste and its management. Such a shift is consequential. For the researchers at the Bureau of Mines, electronic waste as such did not exist. What did exist instead was "electronic scrap" (Dannenberg, Maurice, and Potter 1972, 1). What was of concern was the potential for that scrap to yield secondary precious and base metals (Dannenberg, Maurice, and Potter 1972, 1). Dannenberg and co-workers (1972) do not discuss what might happen to electronic scrap (e.g., landfilling) or why one might be concerned about it were it not recovered by the processes those researchers investigated. The issue of electronic scrap in their view had to do with resource recovery. In contrast, in Arensman's (2000, 110) article, "electronic waste" was a problem because it was "piling up in landfills" and because it was a source of "potentially hazardous substances such as lead, cadmium and mercury, which are currently used in many electronics devices." The contrast illustrates a shift in the framing of discarded electronics: as scrap that is a potential source of resources or as a waste product that must be managed. Of course, these two realities of discarded electronics are not mutually exclusive, but highlighting them underscores the normative effects built into any conception of what discarded electronics are and how they ought to be managed. As historically oriented analyses show so clearly, nothing is "merely" waste; it is made to be so (e.g., Foote and Mazzolini 2012; Strasser 1999; Zimring 2005).

Two years after the publication of Arensman's article an environmental nongovernmental organization (ENGO), the Basel Action Network (BAN), released a report titled *Exporting Harm* (Puckett et al. 2002). The BAN report has since become among the most cited documents in the literature on e-waste, scholarly and otherwise. According to Scopus, a scholarly indexing database, the report has been cited more than 270 times. Were it a peer-reviewed article this citation count would make it the sixth most cited document of any of the more than 2,800 documents with the word "e-waste" or "electronic waste" in their title, abstract, or keywords published since 1973. *Exporting Harm* points to a quite different geography of e-waste than either Arensman's article or the U.S. Bureau of Mines suggests. Instead of "scrap" or "waste" sourced and processed domestically, in the BAN report the story of e-waste is about exports to developing countries.

With the scenarios described above in mind, in this book I explore a deceptively simple question: what is the right thing to do with electronic waste (e-waste)? To answer that question it is first necessary to start afresh with understanding e-waste. The familiar framing of the e-waste problem naturalizes it as being about waste generated in "developed" countries flowing to "developing" countries. In that story line, particular places such as Agbogbloshie, a section of Ghana's capital city, Accra, are said to receive "hundreds of millions of tons" of e-waste from the West (Klein 2009). Such framings of the e-waste problem need to be made strange again because they are part of the problem rather than routes to equitable solutions.

Defamiliarization is a literary technique (Bell, Blythe, and Sengers 2005) that is central to science fiction but is not limited to that genre. It is employed by writers looking to make the familiar strange again for their audiences, often for allegorical effect. Here is a short example of what I mean. The story from which the "hundreds of millions of tons" figure is quoted won an Emmy Award for investigative journalism (*UBC News* 2010). Yet careful examination of the claim that hundreds of millions of tons of e-waste arrive at Agbogbloshie from the West raises serious questions about its veracity. Scholars (e.g., Grant and Oteng-Ababio 2012, 2), citing Klein's (2009) journalism as support, contend that three hundred to six hundred forty-foot containers of e-waste arrive in Ghana per month. But even six hundred such containers arriving monthly will not result in hundreds of millions of tons of e-waste reaching Ghana annually. A standard twenty- or forty foot container can carry a maximum load of 30.48 metric tons (International Organization for Standards 2013). Six hundred such containers arriving per month could, at the most, result in about 219,456 metric tons of e-waste reaching Ghana annually—an amount three orders of magnitude less than the hundreds of millions of tons claimed by Klein (see Lepawsky, Goldstein, and Schulz 2015 for further discussion). World Bank data show that in 2009, the year Klein's documentary aired, Ghanaian ports moved a little over 557,323 twenty-foot equivalent units (TEUs) of container traffic (World Bank 2017). If all those TEUs were devoted solely to carrying electronic waste and no amount of any other commodity into Ghanaian ports, then the maximum tonnage of e-waste entering the country would be just over 16,987,211 metric tons—still an order of magnitude less than hundreds of millions of tons. Of course, formal trade data will miss an unknowable amount of illegal, unrecorded trade. However, the degree to which the hundreds of millions of tons figure ruptures the bounds of plausibility is indicative of why it is important to make familiar stories about e-waste strange again. The anthropologist Carolyn Nordstrom, who has spent years mapping the intermingling of global licit and illicit trade, notes that national economies in Africa show a surprising commonality with those of Italy, Russia, and Peru: at least half the economies of these states "run outside formal reckoning" (Nordstrom 2007, 109); indeed, Nordstrom finds that undeclared transfers of commodities at ports can reach as high 90 percent of what is officially recorded (2007, 126). In other words, at least 50 percent of these economies (and others) goes unrecorded in official trade statistics. Yet to accumulate hundreds of millions of tons of e-waste annually at Agbogbloshie would require the assumption that unrecorded container trade for Ghana is bigger than recorded trade not by 50 percent or even by 90 percent but by more than 400 percent. It is not that the figure of hundreds of millions of tons flowing into Agbogbloshie cited in this award-winning documentary simply overestimates an otherwise plausible metric. It stretches the bounds of plausibility well beyond the breaking point. Yet no correction to the Emmy Award–winning story has ever been issued, and the story continues to be cited as authoritative evidence of how e-waste is dumped in countries of the global south (e.g., Basel Action Network 2016a; Fosdick 2012). Similar stories about the place continue to proliferate (e.g., Hirsch 2013; McElvaney 2014; Souppouris 2016; for important scholarly challenges to conventional thinking about the hazardous waste trade more generally, see Montgomery 1995; O'Neill 2000; Strohm 1993).

E-waste is allegorical. INTERPOL, the international police organization, ran a recently concluded enforcement program against the international trade of e-waste and other hazardous wastes to Africa called Project Eden (INTERPOL 2015). Media exposés of Agbogbloshie, Ghana—a recurrent site of interest in controversies about e-waste—repeatedly describe the place in the biblical iconography of hell and damnation (e.g., Höges 2009; Kaplan 2014; Mongabay 2012; SBS Australia and Vitola 2011). Yet to the extent that the curation of such images ignores the urban political economy of land struggles at the site, the biblical tones of the stories have far more to say about those fashioning them than about the role of e-waste at the site (see Lepawsky and Akese 2015). Like stories about other forms of waste, stories about e-waste index visions of how the world ought to be.

Of course, claims about how the world ought to be raise the questions of whose visions are being articulated, where, when, and under what conditions. I use the term "worlding" to capture the action that generates the dynamics between *is* and *ought*. As I use it, "worlding" refers to two things:

practices that bring together a jumble of not necessarily like things—people as citizens, consumers, and corporations; materials such as plastics, metals, and glass; energy and information; sites and situations that become connected through the use and disposal of digital technologies—as if they cohered to form a common world, and at the same time the research practices that seek to map out and follow the action of those doing the worlding. Research practices are not innocent reflections on an extant world "out there." They too are part of the work of composing common worlds.

How we world a problem like e-waste matters. Particular worldings make some proposals for solutions thinkable and, at least potentially, actionable. At the same time, specific worldings mean other possibilities are not just taken off the table, they are rendered unimaginable, let alone actionable. In the words of the feminist science and technology studies (STS) scholar Maria Puig de la Bellacasa, "Ways of studying and representing can have world-making effects" (Puig de la Bellacasa 2011, 86). In learning how such patterns and processes are organized, it is also possible to recognize that they could be organized otherwise. That is, other worldings are possible, even if achieving those other worldings may not be easy (Sismondo 2017; Woolgar 2011).

Stories about e-waste, particularly those crafted by advocacy groups and the media, are framed so as to inspire devotion strong enough that "it moves an audience to action" (Pine and Liboiron 2015, 3153). In this sense, e-waste might be thought of as what Liboiron (2015) calls a charismatic form of contemporary waste. Arguably, what makes e-waste charismatic is its capacity to act as an allegory of contemporary environmental crises. Contemporary wastes are thoroughly unlike the "dirt" of Mary Douglas's (1966) classic anthropological analysis of purity and taboo. Whereas dirt in Douglas's sense of the term can be managed through ritual and symbolic work with matter out of place, contemporary waste materials defy the efficacy of such work. Contemporary waste materials are "synthetic, unpredictable, and above all heterogenous" (MacBride 2012, 174). They entrain forms of toxicity, heterogeneity, and persistence that mean management practices framed in terms of "cleanup" will never achieve success (Gray-Cosgrove, Liboiron, and Lepawsky 2015).

MacBride (2012) argues that the factors that distinguish contemporary wastes from their historical precedents are their tonnage, toxicity, and heterogeneity. The tonnage of waste arising is enormous. Canada's national statistics agency reports several categories of solid waste, including oil sands tailings (645 million metric tons), mine waste rock (256 million metric tons), mine tailings (217 million metric tons), livestock manure (181 million metric tons), and municipal solid waste (MSW, 34 million metric tons) (Statistics Canada 2012, 11). No directly comparable data exist for the United States or for Europe. However, MacBride (2012, 242) provides data for the annual generation of coal combustion waste (128,000,000 U.S. tons), hazardous industrial wastes (39,580,000 U.S. tons), household hazardous wastes (1,600,000 U.S. tons), construction and demolition debris (330,000,000 U.S. tons), and MSW (249,610,000 U.S. tons). Dizzying numbers like these are widely cited, but they are also deeply problematic. As MacBride (2012) shows, figures for the United States not only are based on very dated studies, they omit entirely important sectors (e.g., manufacturing) and measures other than weight (e.g., toxicity) that are crucial to figure out. Yet doing so is no simple task. For example, the U.S. Environmental Protection Agency has no mandate to track nonhazardous waste generated in manufacturing, so those data are not even collected (see also Liboiron 2016). Similarly, Canadian statistics pay attention to only three primary industries—oil sands production, mining, and agriculture (and only one measure of livestock agriculture at that)—and say nothing at all about manufacturing. We are left with deep uncertainty about those sectors that generate the vast bulk of waste (primary industry and manufacturing). Yet the data we do have strongly suggest that the tonnages of waste for which we have relatively good data, household waste and MSW, are a vanishingly small proportion of overall waste. A tally of the available figures for Canada, the United States, and Europe indicates that household waste or MSW amounts to somewhere between 2 and 9 percent of all waste created. The rest, some 91–98 percent, arises upstream in primary industry, manufacturing, distribution, and retail.

It is crucial to notice this invisibility of waste. When our notion of what waste is and where it comes from is so strongly informed by what we think we know about waste because we have direct daily experience with it through the bins in our homes or the cans we put out on the curb, we have a very partial knowledge of waste. The partiality of that knowledge has practical consequences for how we might imagine solutions to waste problems, such as e-waste, even when we can see them. If you live in

Canada, the United States, or Europe, somewhere between 91 and 98 percent of waste arises before you and I as consumers purchase the products we will eventually discard. When we focus on MSW or postconsumer waste, then we are construing waste as a very particular and partial problem. If we then propose solutions such as individual-, household-, or even municipal-scale recycling, we will be dealing with only a tiny portion of waste.

What we count as waste and how we count it are important issues in other ways as well. It is common to report waste statistics in per capita terms in an attempt to make different places and populations ostensibly comparable. The United Nations University's Solve the E-waste Problem program (UN STEP, a group I am involved with) makes these kinds of statistics available for e-waste generation in countries around the world (United Nations STEP Secretariat 2014). One might quibble with the way those statistics are generated, for example by taking issue with the assumptions that such data reporting activity must make (e.g., relying on estimates of the useful life of electronics before they enter the waste stream). But there is a more fundamental point here. When waste arising is counted in distinct categories, such as MSW, separate from the waste arising from industrial processes and then aggregated and attributed to populations (e.g., of a city, region, or country), then the responsibilities for waste also become attributable to actors, such as individual consumers or households, who have little or no control over waste arising from such activities as mining, design, manufacturing, and distribution—all of which happen before consumers even purchase their devices and which are responsible for the vast proportion of waste generation (Liboiron 2013b). If you purchase a new smart phone, you can choose between a variety of brands and an array of model options. But your choices about the materials used and the labor conditions employed to build that phone are so limited as to be almost nonexistent (though one could point to Fairphone and Puzzle-Phone as two possible partial exceptions). Linking waste and consumer choice in this scenario is a poor way to address the problem at hand.

Contemporary wastes are not just voluminous, they are toxic and heterogeneous. They contain synthetic chemical compounds that do not occur outside the practices of technoscience. Under specific conditions these chemical compounds can cause harm to bodies and ecologies. Electronic equipment is an admixture of material types and compositions.

Their constituent plastics, metals, and glass come in a variety of chemical compositions and physical arrangements with varying potential for harm (this is one reason why electronic items can be difficult and costly to recycle). But where, when, by whom, and under what conditions such harm is experienced are crucial questions to explore when trying to solve the central question this book explores: *What is the right thing to do with electronic waste?* Discards, remainders, and waste from digital technologies arise ubiquitously but unevenly at all points in the production, transport, and use of such items, not merely when they are thrown away. Understanding this uneven ubiquity changes how we might understand e-waste as a problem to be managed and radically alters where efforts should be directed if effective solutions are to be achieved.

Outline of the Book

One of the purposes of this introduction is to start defamiliarizing e-waste. Defamiliarization is an analytical tactic I use throughout the rest of the book. Chapter 2 opens with the May 2015 Conference of the Parties (COP) to the Basel Convention on the Control of Transboundary Movements of Hazardous Wastes and Their Disposal. The Basel Convention was adopted in 1989 and entered into force on May 5, 1992. There are currently 186 parties to the convention, and it is the key international treaty that regulates international shipments of hazardous waste. From its earliest iteration, the Basel Convention has been subject to tumultuous debate. Some coalitions of states and NGOs have been critical from the start about what they perceive as weaknesses of the convention with respect to the enforcement of export prohibitions and environmentally sound management of wastes (see, e.g., Clapp 1994; Myslicki 2009). Other coalitions of states have voiced opposition to what they see as the convention's threat to trade and access to cheaper industrial inputs derived from recycled materials (Kellow 1999). Meanwhile, industrial interests—especially those in the international scrap trade—and their state allies in Europe and the United States have seen the convention as a direct threat to their revenues (Schmidt 1999). As I have discussed elsewhere, the broader background of the convention is important for understanding the international trade and traffic of e-waste (Lepawsky 2015a, 2015b; Lepawsky and McNabb 2010). E-waste meets the Basel Convention's definition of hazardous waste in some circumstances, but does so under conditions of considerable dispute and ambiguity. Moreover, the global patterns of trade and traffic in e-waste (discussed in chapter 4) have evolved such that the regulatory world defined by the Basel Convention is increasingly irrelevant to current and likely future patterns of such trade and traffic.

Among the items debated at the COP meeting were technical guidelines intended to distinguish between "waste and nonwaste" electronics (Basel Secretariat 2014). Those technical guidelines have been debated in draft form since 2010. During the 2015 COP negotiations on these guidelines, the talks collapsed again and the guidelines were adopted in draft form, but not ratified into a legally enforceable provision of the convention. Consequently, there remains no legally precise way to distinguish between discarded electronics that are being shipped as waste (and thus subject to trade prohibitions under the Basel Convention) and those that are being shipped for reuse, repair, refurbishment, or recycling.

Chapter 2 opens up the global trade and traffic of e-waste as a matter of concern. "Matter of concern" is used in a technical way (Latour 2004, 2005) to denote situations in which the solidity of facts is disputed and highly uncertain, especially but not exclusively among the experts themselves. Puig de la Bellacasa (2011) encourages analysts to enrich Latour's notion of matters of concern with a notion of care. "We must ask, 'Who cares?' 'What for?', 'Why do 'we' care?, and mostly, '*How* to care?'" (Puig de la Bellacasa 2011, 96; emphasis in the original). Foregrounding these questions of care is crucial when approaching e-waste as a matter of concern. Typically the problem of e-waste is cast as a postconsumption disposal issue. Yet formatting it that way brackets out all of the practices of resource extraction, design, manufacturing, distribution, and use from which far more waste arises. Postconsumer recycling will never solve the problem of waste arising from electronics production. Even regulations that attempt to reduce or eliminate toxicants in electronics manufacturing raise questions about whose worlds are being cared for and how. Manufacturing might be made less toxic, but doing so is no guarantee that other factors affecting worker safety, such as shift length or the intensity of work on fabrication lines, will also be ameliorated (see Lepawsky 2012).

The chapter follows a methodological approach introduced by Latour and subsequently developed by others (e.g., Marres and Gerlitz 2015; Rekdal 2014; Rogers 2013; Venturini 2010, 2012) to both literally and figuratively map controversies. Mapping e-waste controversies helps make sense of the disagreements that shape e-waste debates. The chapter is intended to help readers make their own judgments about the degree of partiality and situatedness of the competing claims to knowledge about the global trade and traffic of e-waste as advocated by the different actors relevant to the dispute.

The notion of controversy I adopt is derived from exchanges between John Dewey and Walter Lippmann and the conceptual development of their work by later scholars (e.g., Latour 2010; Marres 2005; Venturini 2010). In his famous book, *The Phantom Public*, Walter Lippmann introduced the terminology of "controversies" to refer to particular kinds of intractable problems or disputes. While the specifics of such problems differ, they share the common characteristic of high levels of uncertainty such that even experts disagree on the status of facts relevant to a given controversy. In short, controversies are situations in which "the facts are most obscure, where precedents are lacking, where novelty and confusion pervade everything" (Lippmann 1925, 131). It is when confronting these most difficult of problems that the public is compelled to make decisions about what is to be done (e.g., What is the right thing to do with electronic waste?). The key difficulty for members of such a public—because it is comprised of a diverse collection of people who may or may not have any expertise specific to the situation—is to know which actors in a controversy deserve their support. Over the past decade, a group of scholars has revisited Lippmann's and Dewey's work on public controversies and identified a resource for social analysis in the very difficulties such controversies present (e.g., Marres 2005; Marres and Gerlitz 2015; Rogers 2013). It is in the specifics of such controversies that broader processes of social ordering can be observed in the making (Venturini 2010, 2012).

Chapter 3 examines the legal proceedings of two court cases to explore how a jumble of disparate people, places, and things coheres into a story about e-waste as a problem of global environmental justice. The cases feature Joseph Benson, a Nigerian citizen, who became the first person to be jailed in the UK for exporting electronic waste. Building on claims made in previous chapters and on the work of STS scholars John Law (2004a) and Annemarie Mol (2002), chapter 3 argues that electronic waste is a noncoherent and multiple entity under the law. It is noncoherent and multiple in that two entities, waste and nonwaste, go under the same name—discarded electronics—a situation that then presents judges and juries with the practical problem of composing a legal judgment as if the world were a well-ordered (or "global") whole. Yet what the conjuncture of laws relevant to the case illustrates is that they build in clashing ways of enacting waste and its other, nonwaste, from which wholeness can be only provisionally and contingently achieved. The chapter makes use of the critical legal studies scholar Mariana Valverde's (2009) concept of the "work of jurisdiction" to show how the legal ordering of discarded electronics as waste and nonwaste is more than a merely technical legal exercise: it is also generative of fundamental spatial, temporal, and moral patterns of social order. Rather than a highly abstract instance of legal theorizing, the regulation of discarded electronics as waste and nonwaste institutes practical limits on public decision making about how that which will become waste is manufactured in the first place (see Gille 2007). The regulation of e-waste demonstrates how the practical consequences for who gets what wastes, how, and under what conditions come to be. That those consequences are *made* (e.g., through legal proceedings) rather than emerging from essential characteristics of objects that are either wastes or nonwastes shows that they could be made differently and organized to achieve different, perhaps more equitable outcomes for different people and places.

In chapter 4 I chart global flows of discarded electronics and try to offer a nuanced description and explanation of those patterns. Since the emergence of e-waste as a matter of public concern in the early 2000s, the prevailing characterization of the issue by activists, journalists, and scholars has continued, almost without exception, to be one of waste dumping by the world's rich onto the world's poor. Nevertheless, evidence accumulating from a variety of sources shows that the prevailing characterization of e-waste needs to be rethought. The goal of the chapter is to reconsider preconceived notions of what e-waste is, where it travels, who works with it, and under what conditions—all questions that are explored in subsequent chapters of the book.

Patterns of trade evident in quantitative trade data do not support the dominant story line about e-waste dumping. Field-based studies in a variety of locations thought to be "hot spots" of e-waste dumping also point to a vastly more nuanced and complex story of growing domestic consumption of electronics and the role of reuse, repair, and refurbishment in those locales of electronics discarded elsewhere. Drawing on a variety of quantitative and qualitative studies, including my own research and fieldwork, chapter 4 highlights how the dominant e-waste story line fails to explain the global patterns of its trade and traffic. The chapter points to the role of trade for reuse and repair of electronics, particularly where people are less able or unable to afford newly manufactured devices. As such, the chapter illuminates some of the worlds of reuse and repair that are crucial for a fuller understanding of the global trade and traffic in electronic waste. It argues that a better explanation for the historical and emerging patterns of trade and traffic in discarded electronics lies not in trade for dumping but in trade for reuse and the dynamic ecologies of repair, refurbishment, and materials recovery that drive that trade. The argument is not that disposal of end-of-life electronics or the toxicological consequences thereof never occur. Instead, the argument is that when disposal does occur, it is typically after multiple rounds of reuse, repair, refurbishment, and materials recovery in which domestic consumption, rather than foreign dumping, plays a crucial role. As a consequence, the character, distribution, and scale of harms and benefits arising from global flows of e-waste cannot be sufficiently understood if the dominant story line about waste dumping remains the unquestioned starting point for specifying discarded electronics as a waste management problem.

Chapter 5 asks how we know electronic waste. While this question might seem to have been a logical place to begin the book, it was more pressing first to challenge prevailing story lines about e-waste (and waste more generally) before addressing this deceptively simple query. Waste in general, and e-waste specifically, are notoriously difficult to quantify (MacBride 2012). Most available data on waste generation are collected at the individual or household level and are counted as MSW. MSW data tell us little about waste arising in industrial production. The problem of knowing waste is partly a problem of what to count and where. For example, data on e-waste generation typically are measured in terms of weight. Weight is relatively easy to measure, but it tells us nothing about other important characteristics of e-waste, such as its potential for toxicity (e.g., 1 kilogram of aluminum and 1 kilogram of mercury are identical in terms of weight but radically different in terms of toxicity). When e-waste generation is measured as what individuals, households, or businesses put into the municipal waste stream, the vastly larger amounts of solid, liquid, and gaseous "externalities" arising from electronics manufacturing are bracketed out of consideration of what constitutes waste from electronics. Thus, to measure waste in some ways and not others is a problem of knowledge or epistemology—how we come to know what we think we know—with practical consequences: to construe e-waste as a postconsumer waste management problem suggests that practical policies such as household recycling are the right solution. Yet if most of the waste arising from electronics occurs before consumers even purchase them (as happens in raw materials extraction and manufacturing), then household recycling programs will do little to mitigate waste generated from electronics.

Chapter 5 makes use of a movement within the STS literature sometimes referred to as the "social life of methods" (Law 2004b, 2010; see also Barad 2007). The STS literature argues that methods cannot be understood as neutral tools for observing the world. Instead, methods are partly generative of the phenomena they purport to study. Methods have a "double social life" because they are articulated by and from particular places and people and because methods have generative consequences—they partially create the very situations they investigate. The chapter explores the implications of the double social lives of methods for studying e-waste. In particular, the chapter examines how various methods for researching and generating knowledge about e-waste are fashioned into analytical narratives. The chapter pays attention to specific methods as knowledge-making practices. In particular, I examine the implications of the use of asset tag and photographic evidence to tell stories about the origin of e-waste found in sites in the global south. Asset tags, photography, and certain methods used to collect them tell us little about the various "afterlives" of those electronics and their potential circulation through multiple preceding rounds of repair and reuse. On the other hand, the quantitative trade data available for e-waste and used by some scholars (including myself; see Lepawsky 2015b) and policy makers offer a global synopsis of flows, but of only a very narrow range of discarded electronics, and by definition such data miss unreported trade and illicit trade. The chapter investigates what implications the social life of methods has in terms of how the analytical results of those methods are taken up in subsequent attempts to intervene in the practical management of global flows of e-waste.

Chapter 6 opens with a seemingly quirky question: *How much does the Internet weigh?* There is still a tendency to assume the information age is built on an ethereal, virtual, and weightless network. Any thought given to the material consequences of electronics is typically limited to what happens before they are manufactured (e.g., in the mining of a specific class of raw materials sometimes called "conflict minerals") or after they are thrown away as waste (for an important exception, see Smith, Sonnenfeld, and Pellow 2006). The chapter playfully opens up to scrutiny this pernicious tendency.

The point of the chapter is twofold. The first is to continue the analytical approach of defamiliarizing digital technologies and discards. Chapter 6 documents some of the geographies of matter, energy, and labor needed for the creation and maintenance of digital technology and information networks so that they may run smoothly. It charts certain kinds of scenes, or what I call the "minescapes," "productionscapes," and "clickscapes" of electronics. I refer to the wastes arising from these scenes collectively as the "discardscapes" of electronics. These discardscapes are a kind of archipelago—patchy, uneven, and not necessarily coherent. The analogy with an archipelago points to the second aim of the chapter: to document the degree to which discards, remainders, and waste from digital technologies arise unevenly but at all points of their existence, not merely if and when they are thrown away.

Documenting the scale and location of waste arising from digital technologies throughout their existence is done for two reasons: first, to demonstrate that resource extraction for and the manufacturing of electronics generate vastly more waste than does the postconsumption discarding of gadgets; and second, to more fully develop a point raised in earlier chapters, that the dominant story line about e-waste being a postconsumer waste management problem is a partial, even peculiar, framing of the problem of e-waste since it typically leaves out the vastly higher amounts of waste arising from resource extraction and manufacturing necessary for the creation and use of digital technologies. The importance of the latter point is further developed in this chapter by noting how bracketing out waste arising in resource extraction, manufacturing, and use guarantees that the e-waste problem cannot be solved through postconsumer recycling. Yet it is just such recycling that is *the* prevailing policy prescription for managing e-waste, and one that derives from the dominant story line about it.

Drawing on the work of Liboiron (2009) and MacBride (2012), chapter 6 argues that since the vast bulk of waste is generated before devices reach consumers, the hope that recycling electronics will mitigate e-waste is a false one. In these ways, the chapter deepens the analytical work of defamiliarizing electronic discards. It also plays on the concept of "weightiness" in the double sense of the massive energy and material consequences of waste arising in resource extraction, manufacturing, use, and disposal of electronics and of the seriousness and importance of those consequences in their highly uneven distributions among people and places.

Chapter 7 draws the book to a close by returning to its central question—what is the right thing to do with electronic waste?—and critically examining a number of possible solutions for mitigating waste arising from the manufacture, use, and disposal of electronic items. Some of these solutions, particularly recycling and trade prohibitions, already exist and are gaining in popularity among policy makers. Other solutions, however, like other worlds, are possible. These other solutions include both more and less conventional possibilities. Among them would be decriminalizing export for reuse, repair, and elective upgrade; facilitating ethical trade in electronics reuse, repair, refurbishment, and recycling; implementing genuine extended producer responsibility, which would force brand makers to internalize the costs of externalities and waste management; and democratizing industry by instituting robust forms of public oversight over how much and what kind of wastes can be produced at all.

The chapter explores the possibilities of and limits to several concrete solutions. These include changing regulations to increase the likelihood that "clean(er)" and "green(er)" electronics are manufactured by looking to a regulatory model that operates as a norm in other multibillion-dollar industries (e.g., foodstuffs and pharmaceuticals) and adapting it for another multibillion-dollar industry (electronics). Pharmaceutical firms, for example, must demonstrate to the U.S. Food and Drug Administration that their products will not cause unacceptable harm *before* those products are manufactured for the consumer market. Why should not electronics manufacturers be subject to a similar obligation? This possible solution is, of course, highly speculative (and would no doubt be very contentious were it even to be broached by policy makers), but the chapter uses this speculative suggestion to extend the analytical tactic of defamiliarization central to the book's argument. In effect, the chapter asks readers to question the assumptions they may have that recycling will solve the e-waste problem and open themselves to imagining other possible worldings of electronic waste that may have more desirable consequences by asking themselves whose worlds are made cleaner and greener, where, how, and under what conditions when solutions to e-waste are proposed and instituted.

From this discussion of conventional and unconventional solutions to e-waste, the chapter closes on a note of invitation to adopt as a framework for thinking about the problem three broad propositions that arise from the empirical and conceptual material addressed throughout the book. When attempting to answer a question about waste, such as *What is the right thing to do with electronic waste?*, it is important to keep in mind that hazard and waste are indeterminate; that waste can be noncoherent; and that we must defamiliarize waste or, said differently, make waste strange again so as to keep any proposed solutions open to critical reflection about their efficacy for mitigating waste as such.

2 Waste/Nonwaste

The difficulty of controversy is not that actors disagree on answers, but that they cannot even agree on the questions.

—Tomasso Venturini (2010, 262)

Introduction

In May 2015, some 1,200 people gathered for the 12th Conference of the Parties (COP 12) to the Basel Convention on the Control of Transboundary Movements of Hazardous Wastes and Their Disposal, in Geneva, Switzerland. "From science to action, working for a safer tomorrow" was the event's tag line (Basel Secretariat 2015d). Among the outcomes highlighted on the Basel Secretariat's postconference web page is the "adoption of the e-waste technical guidelines" (Basel Secretariat 2015d). The anodyne language of the announcement belies a highly disputatious unfolding of events leading up to and including their adoption at the May meeting.

The guidelines' somewhat unwieldy full title is *Technical Guidelines on Transboundary Movements of Electronic and Electrical Waste and Used Electrical and Electronic Equipment, in Particular Regarding the Distinction between Waste and Non-Waste under the Basel Convention*. As the title suggests, the guidelines are intended to provide regulatory clarity for authorities (e.g., customs agents at ports) who are trying to determine whether an export shipment is electronic waste and thus subject to control under the Basel Convention.

The technical guidelines had been debated in draft form since 2010. At the May 2015 COP meeting, negotiations around these guidelines collapsed and the guidelines were adopted only on an interim and non–legally binding basis. Documents from the meeting show several representatives were unable to support adoption of the guidelines after last-minute revisions were made. At least one delegate felt the draft was so full of loopholes it amounted to "Swiss cheese" (Basel Secretariat 2015a, response by BAN, p. 1). Another delegate, incensed by the manner in which the deliberations over the guidelines had proceeded, announced that her delegation "wished to dissociate itself from the proceedings" (Basel Secretariat 2015b, 17). The adoption of the guidelines under such conflictual and provisional circumstances means there remains no legally precise way under the Basel Convention to distinguish between discarded electronics that are being shipped as waste (and thus subject to trade prohibitions under the Basel Convention) and those that are being shipped as nonwaste for other purposes, such as reuse, repair, or refurbishment. How can such a seemingly basic distinction—that between waste and nonwaste—be so intractable?

In this chapter I follow the practice of controversy mapping to chart some answers to this question. Like actual cartographers, those of us who are not already experts on a given issue are thrown into the middle of things when they erupt into public disputes, registered in our newspapers or newsfeeds. When such controversies do erupt we almost always know little to nothing about the internal details. Like cartographers, we have to find some ways to orient ourselves without full knowledge of the terrain, who the participants are, what they stand for, or who deserves our trust.

Controversy mapping involves analyzing social debates, particularly but not exclusively those involving issues of public concern generated by science and technology (Marres and Gerlitz 2015; Rogers, Sánchez-Querubín, and Kil 2015; Venturini 2010). Among practitioners, the terms "controversy" and "issue" have technical meanings. A controversy is taken very broadly to be anything over which there is public disagreement, but it is not disagreement per se that signals a genuine controversy. Instead, what does is that actors may not even agree on the framing of the question. To paraphrase Venturini (2010, 262), ask a simple question, such as "Is a discarded cell phone waste or not?," and the actors will ask where and in what jurisdiction the phone was discarded. They will begin arguing about whether it is functional in full or in part, and if in part, which parts. They will dispute whether it can be repaired on technical grounds, on economic grounds, whether it is under warranty, and the conditions under which it might be reusable. In other words, they disagree not only on the answers but on the questions as well.

Controversy mapping is an approach to charting terrains of disagreement and actor partisanship. Partisanship sometimes has a pejorative connotation, but my use of the term is restricted to its denotative range of meaning without implying disapproval. As I use it, partisanship signals two intertwined ideas. The first idea is that all knowledge that informs an actor's position is partial and situated, built as it is on more or less trustworthy but always incomplete claims and evidence. The second is that since everyone's position (including my own) on issues is partial, situated, and constructed, there is no binary of partisan and nonpartisan; there is only a continuum of partisanship. However, that all knowledge informing a position is partial and situated (i.e., partisan) does not mean that all knowledge and positions are equally so. Partisanship covers a very wide latitude of positions. Its expressions include everything from being a supporter, fan, or champion to being a stalwart, zealot, or fanatic. There are very large differences between being a supporter and being a fanatic. A supporter of a given position might be inclined to modify her position when new evidence becomes available. Zealots and fanatics, on the other hand, tend to stick to their guns. Detecting differences in partisanship is particularly germane to navigating controversies. Coming to grips with the specifics of disagreements with respect to the technical guidelines also provides a set of waypoints to assist in navigating the complexities of e-waste addressed later in the book.

What makes controversies and their issues of interest is that their emergence and persistence make it possible to sketch out the convoluted action by which collective life is generated. Analysis entails the detailed and methodical examination of the elements of a phenomenon for discussion and interpretation. It is not synonymous with explanation—an account of causes—although explanation may be one of the goals of analysis. Venturini (2010) likens analyses premised on controversy mapping to swimming in magma. What he means by this comparison is that controversies provide occasions for a gamut of actors to stake out positions, alliances, and oppositions such that for the analyst, the work of making collective life gel becomes traceable. Geologically speaking, magma is both solid and liquid. As a metaphor for the social-in-the-making, magma implies this kind of both/and action of collective life taking shape if and when it cools and solidifies. The significance of controversies, however, is that they help clarify that the particular shape that collective life may eventually take is built and, having been built, could be assembled differently. Swimming in the magma of controversy, in other words, helps one sense the action of building collective life before the social cools down and solidifies, becoming taken for granted, as merely "the way it is."

During controversies, part of the action includes actors articulating not just how the world is but how it ought to be. Proponents of controversy mapping refer to the idealized world that an actor and its allies are trying to institute as a "cosmos." A cosmos is a vision of the world as a well-ordered whole. As such it is an ideal, that is, a vision of how the world ought to be even if it is not actually that way. Not all of the world visions that emerge during a controversy are mutually exclusive, but some are. Where cosmoses conflict on mutually exclusive grounds, questions of whose worlding will prevail come to the fore. These are political questions—questions pertaining to struggles over the power to institute this or that worlding. "Can we live together?" the actors might ask of one another's positions (Latour 2005, 254). When and where the answer is no signals instances in which the magma of controversy fails to congeal into the terrain of a common world, the controversy carries on, and a cosmos shared by disputatious actors fails to materialize. As I show later, the positions of actors in the debates over the technical guidelines give rise to different cosmoses, some versions of which are mutually incompatible. That incompatibility is a key reason why negotiations broke down at COP 12 and why there remains no legally precise way to distinguish waste from nonwaste electronics under the Basel Convention.

To develop my analysis of the guidelines controversy I draw on the classic work of Brian Wynne (1987, 1989) on hazardous waste regulation. My reason for doing so is that the drama leading up to and including COP 12 played out in miniature a script predicted almost thirty years earlier in Wynne's writing. As Wynne (1987, 11) shows, science tends to be scripted into the regulatory process for hazardous wastes in a dual role. One of those roles is to build an empirical understanding of risks. The other role is more symbolic. Science acts as a guarantor of public trust in, and the credibility of, regulations and regulators. When it comes to hazardous waste, however, science is poorly suited to playing these roles. This is because, Wynne says, the meanings of "hazard" and "waste" are "not just imprecise or statistically fuzzy—they are *fundamentally* ambiguous" (Wynne 1987, 8;

emphasis in the original). The meanings that "hazard" and "waste" do take on cannot be fully grounded in a scientific understanding of the physical properties of materials alone. Instead, their meanings are also inextricably tangled up with specific institutional arrangements designed by people. Those arrangements tend to operate in ways that do not conform to quantifiable measures of uncertainty for risk analysis. One example discussed by Wynne (1987, 36) is the differential regulation of the same chemical compounds, polychlorinated biphenyls (PCBs), in the United States and the United Kingdom. The former country permits a concentration limit of 50 ppm for disposal in landfills. The latter permits concentrations of no more than 10 ppm. There are no exclusively physical chemical criteria by which to judge one of these thresholds as "right" and the other as "wrong" or to call one "hazardous" and the other "not hazardous." They are both/ and, yet differ by a factor of five.

Wynne (1987) goes on to develop a key distinction between uncertainty and indeterminacy as they pertain to questions of hazardous waste (see also Hird 2012). Uncertainty is, in principle, knowable within a range of possible values. By definition, indeterminacy is not. As Wynne shows, it is indeterminacy rather than uncertainty that most fundamentally characterizes questions of hazardous waste. The indeterminacy of hazardous waste arises for several reasons. One is the thresholds used to define hazardousness, as in the U.S.-UK example of PCBs. But regulatory approaches to hazardous waste also typically distinguish it from other categories of waste, such as household waste or municipal waste. However, some household products, such as batteries and cleaning fluids, that make their way into municipal waste and sewage infrastructure meet or exceed thresholds of toxicity otherwise used to define hazardous wastes. Yet those thresholds are often relaxed in regulations, sometimes by orders of magnitude, lest all household waste and sewage have to be treated (very much more expensively) as hazardous waste.

Furthermore, we lack the capacity even in principle to assess all extant and new chemicals that annually become available for industrial uses, even if such testing were mandated by regulation (Wynne 1987, 48; see also Hird 2012, 457). At the time of Wynne's research, roughly seven million chemicals were known to exist, of which 80,000 were in commercial use, with a thousand new chemicals becoming available annually. Even if all testing capacity on Earth were devoted to assessing toxicity, only about five hundred such chemicals could be tested for their capacity for harm (e.g., carcinogenesis). The overwhelming of testing capacity largely remains the case today as at the time of Wynne's research. In other words, with respect to the hazardousness of the chemical constituency of that which will eventually become waste, there are fundamental limits to what we can know even in principle. It is not that we do not know what hazards might await in the expanding chemical galaxy we inhabit but that we cannot know. This is the difference between uncertainty and indeterminacy. Uncertainty implies knowability within a range of probabilistically defined possibilities: Q can be as much as X and as little as Y, no more and no less. Indeterminacy is different. It registers a fundamentally unfixed foundation on which to build knowledge. Hazardous waste in this sense is a nonfoundationalist entity—it has no solid ground floor that makes hazardous waste *hazardous waste* always, forever, and everywhere (see also Wynne 1989).

Moving from a Statement ...

The Conference of the Parties

1. Adopts, on an interim basis, the technical guidelines on transboundary movements of electrical and electronic waste and used electrical and electronic equipment, in particular regarding the distinction between waste and non-waste under the Basel Convention, on the understanding that the technical guidelines are of a non–legally binding nature and that the national legislation of a party prevails over the guidance provided within the technical guidelines. (Basel Secretariat 2015a, 1)

The technical guidelines are part of the Basel Convention on the Control of Transboundary Movements of Hazardous Wastes and Their Disposal (1989), a cornerstone of international governance of the trade of toxicants. The Basel Convention entered into force on May 5, 1992 (Kummer Peiry 1995). Key parts of the convention that are relevant to e-waste remain sources of disagreement among the parties to the convention. Annexes VII, VIII, and IX are of particular relevance. Annex VII signatories are the member states of the Organisation for Economic Co-operation and Development (OECD), the European Community, and Liechtenstein. Non–Annex VII signatories are all other signatories to the convention. Annex VII is pertinent in its relation to an amendment of the convention adopted in 1995, known as Decision III/1 or the "Ban Amendment" (Basel Secretariat 2017). That amendment prohibits Annex VII signatories from trading to non–Annex VII signatories. However, the amendment has not yet been ratified by enough signatories to the convention to enter into force.

Annexes VIII and XI respectively define specific lists of materials that are deemed to be hazardous waste (and thus subject to Basel Convention control) or not waste (and thus outside the prohibitions of the convention). However, both Annex VIII and Annex XI enumerate a wide range of materials that pertain to electronics. As a consequence, these materials have an ambiguous status as both waste and nonwaste under the convention (see table 2.1).

Table 2.1

List of waste and nonwaste electronics in Annexes VIII and IX of the Basel Convention

Annex VIII, "List A"	Description	Annex IX, "List B"	Description
A1160	Waste lead-acid batteries, whole or crushed	B1090	Waste batteries conforming to a specification, excluding those made with lead, cadmium, or mercury
A1170	Unsorted waste batteries, excluding mixtures of only list B batteries	B1110	Electrical and electronic assemblies: Electronic assemblies consisting only of metals or alloys Waste electrical and electronic assemblies or scrap (including printed circuit boards) not containing components such as accumulators and other batteries included on list A, mercury switches, glass from cathode-ray tubes and other activated glass and PCB capacitors, or not contaminated with Annex I constituents (e.g., cadmium, mercury, lead, polychlorinated biphenyl) or from which these have been removed, to an extent that they do not possess any of the characteristics contained in Annex III (note the related entry on list A A1180) Electrical and electronic assemblies (including printed circuit boards, electronic components and wires) destined for direct reuse, and not for recycling or final disposal

Table 2.1 (continued)

Annex VIII, "List A"	Description	Annex IX, "List B"	Description
A1180	Waste electrical and electronic assemblies or scrap containing components such as accumulators and other batteries included on list A, mercury switches, glass from cathode-ray tubes and other activated glass and PCB capacitors, or contaminated with Annex I constituents (e.g., cadmium, mercury, lead, polychlorinated biphenyl)	B1115	Waste metal cables coated or insulated with plastics, not included on list A A1190, excluding those destined for Annex IVA operations or any other disposal operations involving, at any stage, uncontrolled thermal processes, such as open burning
A1190	Waste metal cables coated or insulated with plastics containing or contaminated with coal tar, PCB, lead, cadmium, other organohalogen compounds		
A2010	Glass waste from cathode ray tubes and other activated glasses		

Source: Basel Secretariat (2011a).

A total of six draft technical guidelines were produced between 2010 and 2015. Each draft is posted online, along with comments received from a variety of actors, some of whom participated in the negotiations over each version of the text and some of whom provided comments on the publicly released drafts of the guidelines. These actors range from delegates of national government agencies to representatives of trade associations, individual businesses (e.g., Philips), and an environmental NGO (ENGO), the Basel Action Network. Altogether approximately ninety-nine documents were submitted by thirty-four individual actors as commentary on drafts of the guidelines. Each draft of the guidelines was informed by a substantial number of positions articulated by actors.

Venturini (2010, 262) recommends focusing analytical attention on "where everyone is shouting and quarreling" to map out key areas of conflict among actors. Doing so helps draw attention to what matters most for the actors themselves. It also provides a helpful selection device: since a great many individual positions are articulated with respect to each draft of the guidelines, it is easy to become lost in their proliferation. Thus, identifying those points on which there is the sharpest disagreement and the positions of actors that are most divergent helps to usefully order the positions into hierarchies of relative importance and thereby aids in coming to grips with the complexities of the debates. To accomplish this analytically, I identified those disagreements evident when an actor (or actors) explicitly articulated a position in opposition to a position taken by another actor (or actors); alternatively, I identified disagreements where the content of stated positions contradicted other position content, even when the actors who articulated these different positions did not explicitly do so in response to the position of another actor. To do so, I collected all publicly available documents on the Basel Convention website that provided comments on each version of the guidelines. I read each document carefully to identify the main points of disagreement, then used a web-based platform called Debate Graph to visualize the relationships between different actors and their positions and disagreements with one another as articulated in the documents, recording their comments on each draft of the guidelines (see figure 2.1 online).

... to Actors and Debates

Some points of contention that emerged in the early drafts quickly fell away as negotiations evolved. One example is the scope of the technical guidelines, which was a subject of debate during early rounds of the negotiations. Scope, however, fairly quickly became subordinate to more contentious points. Other disagreements persisted across multiple drafts of the guidelines, such as how the guidelines would define the difference between waste and hazardous waste or what statistics the guidelines relied on in the preamble, titled "About E-waste," that justified the need for the guidelines. However, of all the myriad issues that arose in the negotiations over the guidelines, only one cut across every draft of them, became the focus of the most intense disagreements, and eventually formed the core of the breakdown of negotiations in May 2015: how to define situations in which used equipment is or is not waste (see figure 2.2 online). In what follows, I pay attention to the issues on which the most disagreement among actors was evident. I then devote substantial attention to the most contentious of these issues, the distinction between used equipment that is waste and that which is nonwaste.

Initially, the scope of the technical guidelines was an issue among negotiators (see figure 2.3 online). While twenty-one positions could be identified in the documents submitted by actors as comments on drafts of the guidelines, only three of these became areas of clearly articulated disagreement among actors. These three areas were (1) whether the guidelines should make reference to the Ban Amendment, (2) whether the guidelines should facilitate trade, and (3) whether the guidelines should incorporate the parallel discussions that were occurring in two other Basel Convention initiatives (one called the Partnership for Action on Computing Equipment, or PACE, and another called the Mobile Phone Partnership Initiative, or MPPI).

An early item of concern was how the technical guidelines would connect with the PACE and MPPI initiatives already under way. Canada, the United States, the European Union, and the Information Technology Industry (ITI) Council, an industry trade association, all suggested that the technical guidelines should harmonize with the extant efforts of the MPPI and PACE (see figure 2.4 online). BAN, an ENGO that has been active in Basel negotiations for more than two decades, opposed such harmonization. The group claimed that the work of MPPI and PACE "ha[s] been heavily weighted toward industry participation and participation by OECD stakeholders" (Basel Secretariat 2011b, p. 1 of response by BAN).

BAN's response implies that industry and OECD stakeholder positions are colored by commercial considerations. As I discuss below, such commercial interests are evident in some positions taken by some industry and other actors. There are, of course, differences between for-profit manufacturers and nonprofit ENGOs. The former operate with massively larger revenues, for example. But if the distinction between "industry" and "ENGO" is premised on one side of the divide having a commercial interest (i.e., industry) in the outcome of the debate and the other not, then the divide is too blunt to effectively assess the partisanship of the actors involved in this controversy. In this case, manufacturers, their trade associations, and BAN all have different commercial interests in the outcome of the technical guidelines. For the manufacturers and their trade associations, those interests are principally about how capital goods, medical devices, and professional equipment will be handled under the guidelines. On the other hand, at around the same time as the first draft of the guidelines was being negotiated, BAN was launching its e-Stewards recycling certification program. The e-Stewards certification includes among its provisions restrictions on trade that mirror those of the Basel Ban Amendment. By 2011 that program accounted for more than 57 percent of the nonprofit's revenues and for more than 76 percent by 2014 (Propublica 2015). Of course, BAN's absolute revenue is tiny compared to the revenues of the companies and trade associations that had input into the drafting of the technical guidelines. Yet BAN relies heavily on revenue from its e-Stewards program to fund its operations. Having the guidelines align with BAN's e-Stewards program would strengthen the certification's brand value in the marketplace (Jones 2011, 3). Thus absence of a financial stake in the outcomes of these negotiations is not a characteristic that can be used to distinguish the partisanship of actors with respect to a given position.

Whether the guidelines should refer to the Ban Amendment arose out of a suggestion by BAN. In its response to the first draft of the technical guidelines, BAN suggested that even though the Ban Amendment was not yet in force, its "implementation by many countries has led to numerous national prohibitions" (Basel Secretariat 2010, p. 6 of response by BAN) against importing hazardous waste from Annex VII signatories (the OECD, European Community, and Liechtenstein) to non–Annex VII signatories (all other parties). The United States articulated a directly opposite view, arguing that "a reference to the Ban Amendment should not be included" in the technical guidelines (Basel Secretariat 2012b, p. 8 of U.S. response; see figure 2.5 online).

The United States has signed but not ratified the Basel Convention (making it the only OECD country not to do so). The decision not to ratify the treaty is a response to perceived conflicts between the Basel Convention and the United States' own Resource Conservation and Recovery Act (RCRA). A U.S.- based industrial lobby concerned about how the Basel Convention could impinge on trade in recyclable materials also played an important role in the U.S. decision not to ratify the treaty (Schmidt 1999). This issue of trade worked its way into the U.S. position with respect to the technical guidelines, too.

Commenting on the scope of both the May 2012 and September 2012 drafts of the technical guidelines, the United States articulated a pro-trade position (see figure 2.6 online). Advocating for a voluntary procedure of notification for trade transactions destined for repair or refurbishment, the United States suggested that such a procedure would "help facilitate legitimate commercial transactions" (Basel Secretariat 2012a, p. 4 of U.S. response). The U.S. position on shaping the scope of the guidelines so as to facilitate trade was directly opposed by both BAN and Colombia. BAN argued that "trade is not inherently good," pointing out that "if equipment is moved to locations that lack collection and recycling infrastructure you are dooming areas of the world to a serious hazardous waste legacy from discarded e-waste" (Basel Secretariat 2012a, p. 5 of response by BAN). Similarly, Colombia articulated a counter argument to the U.S. position, claiming that "the purpose of the guidance is not to facilitate the trade, but to clarify the control procedures for the transboundary movements of used equipment and e-waste" (Basel Secretariat 2012a, p. 1 of response by Colombia).

The question of export for repair became an important point of contention within debates over the scope of the technical guidelines and how they would appropriately balance prohibiting waste flows and permitting legitimate commercial activity to continue (see figure 2.7 online). ITI, a trade association for the IT industries, suggested that the guidelines explicitly "recognize the economic, social and environmental benefits that can accrue from the continued use of electrical and electronic equipment, even reuse following repair or refurbishment in the receiving country" (Basel Secretariat 2010, p. 6 of response by ITI). The ITI's position was opposed, however, by both BAN and Norway. BAN argued in part that "very often the export takes place to avoid costs of more diligent environmentally sound management" (Basel Secretariat 2010, p. 5 of response by BAN). Norway worried similarly about the "possibility of using repair centre as a loophole to circumvent the [Basel] Convention" (Basel Secretariat 2010, p. 1 of response by Norway).

BAN would later articulate a position that if export for repair, refurbishment, or reuse activity entailed any waste arising from those activities in non-OECD countries (i.e., non–Annex VII signatories), then the export had to be treated as a waste export. This position would be strongly opposed by IT companies and trade associations, which pointed out that the repair, refurbishment, and reuse of two particular classes of used equipment, professional or capital goods equipment (i.e., enterprise-level rather than consumer-level electronics) and medical devices, are deeply reliant on international reverse logistics chains to support those activities. Such reverse logistics chains, these actors pointed out, already existed and did not map to the neat binary world geography distinguishing Annex VII and non–Annex VII territories from one another. The issue of repair would become a core part of the broader and more contentious issue of how the technical guidelines should handle used equipment.

The distinction between "waste" and "hazardous waste" formed another broad area of debate during the drafting of the technical guidelines. BAN proposed adding a list of hazards in electronic equipment, using two annexes of the Basel Convention, Annex VIII and IX, as guides (see figure 2.8 online). The proposal received support from Norway, Colombia, and Global Electric Electronic Processing (GEEP), an e-waste recycling firm. Norway suggested it would be useful to have a "non-limited list of hazardous components and substances in an appendix to the guideline" (Basel Secretariat 2012a, p. 1 of Norway's response). Colombia expressed a generally supportive position in favor of BAN's proposal but noted that "implementation for making decisions by the authorities on the control of transboundary movements is limited because such information is indicative but not conclusive about the dangerousness of the e-waste" (Basel Secretariat 2012a, p. 3 of Colombia's response). GEEP was more definitively supportive, recommending the guidelines "provide common guidelines within this non-exhaustive annex as to the definition of hazards" (GEEP to Basel Secretariat 2012a, p. 3 of response by GEEP). But the inclusion of such a list was deeply contested in a suite of opposing arguments articulated by Canada and the United States, as well as by Philips (the electronics firm) and ITI. Canada argued the list approach proposed by BAN "would be problematic given that 'e-waste' captures a very broad range of electrical and electronic equipment (EEE) and many countries currently define and characterize e-waste differently under their national laws" (Basel Secretariat 2012a, p. 1 of response by Canada). The United States made two opposing claims against the proposed list: that it contained "a number of inaccurate entries" and that, under the Basel Convention itself, "All listed wastes have the opportunity to be classified as non-hazardous based on a showing that they do not express an Annex III [of the Basel Convention] characteristic" (Basel Secretariat 2012a, p. 1 of the U.S. response). Philips argued that the proposed list was "limited in scope, focusing on computers and displays, which provides minimal to no guidance on a wide variety of other electronic products" and that it would be better to list characteristics of hazards, rather than specific substances of types of equipment (Basel Secretariat 2012a, p. 1 of response by Philips). ITI argued that there was no need for such a list since the Basel Convention already defined hazardous materials in Annexes VIII and IX (see table 2.1 and figure 2.9 online).

Eventually a compromise was reached that relied on the relevant annexes of the Basel Convention. In lieu of a list of type of equipment (e.g., desktop and laptop computers, monitors) along with the locations in a given device where Basel listed materials could be found, the technical guidelines adopted at COP 12 favored deference to Annex VIII and IX of the Basel Convention. While the decision to favor the extant annexes of the Basel Convention quelled the dissent over inclusion of the list of equipment proposed by BAN in the technical guidelines, it left unresolved a crucial area of deep ambiguity that remains in the Basel Convention itself. Annexes VIII and IX respectively define wastes considered hazardous (and thus subject to prohibitions against trade) and those that are not hazardous (and thus outside the scope of the Basel Convention). Both annexes list materials that pertain to electronics. In other words, within the Basel Convention itself various forms of electronics are listed as hazardous waste and as wastes not covered by the convention. Technically, the distinction is between Annex VIII "List A" and Annex IX "List B" (see table 2.1). The ambiguity generated by these lists is magnified in the Basel Convention in at least two ways. The first area of ambiguity is in the definition of "direct reuse." Direct reuse "can include repair, refurbishment or upgrading, but not major reassembly" (Basel Secretariat 2011a, 83 fn. 20), yet what counts as "major reassembly" is not specified. Second, both Annex VIII and Annex IX refer to Annex III of the convention, which specifies hazardous characteristics of materials. While Annex III lists a substantial array of such characteristics, it also notes:

The potential hazards posed by certain types of wastes are not yet fully documented; tests to define quantitatively these hazards do not exist. Further research is necessary in order to develop means to characterize potential hazards posed to man and/or the environment by these wastes. (Basel Secretariat 2011a, 59)

In the absence of definitive tests, it becomes possible for different parties to the convention to make arguments either way for specific materials or classes of materials—pace the argument of the United States about all material having the opportunity to be classified as nonhazardous and thus outside the scope of the convention.

All drafts of the technical guidelines include introductory remarks about the document's scope and the problem they are meant to address. Section B of these introductory remarks, titled "About E-Waste," saw substantive disagreement over claims made about the scale of e-waste generation and transboundary flows. Debate focused especially on a statistic initially attributed to a study by Xiezhi and co-workers (2008).

The first draft of the technical guidelines states, "Xiezhi et al (2008) has suggested that about 50–80% of the E-waste from industrialized countries ends up in South-east Asia for cheap recycling due to the low labour costs and less stringent environmental regulations in this region" (Basel Secretariat 2010, 5). It is this figure of 50–80 percent that became the focus of debate. Chronologically, BAN addressed the figure first in the group's comments on the first draft of the 2011 guidelines. No other actors picked up on the issue until the next iteration of the guidelines. BAN's comments offer important contextualization of the 50–80 percent figure and are worth quoting in full:

I [Jim Puckett, head of BAN] would check this quote [referring to Yu Xiezhi and co-workers' study] or perhaps find another, because it does not take into account the amount that goes into landfills/incinerators nor the amount that is in storage. Also it assumes so much going to South-East Asia, when in our [BAN's] experience the vast majority of wastes directed to Asia goes to South Asia or China. BAN is largely responsible for the 80% figure that is found in the literature today. That figure was introduced in our report Exporting Harm (2002) and came from an industry insider/consultant in the United States who is very knowledgeable about the e-waste marketplace. We needed to use an industry estimate because the data, to this day does not exist. However the figure is for amount of e-waste that after being delivered to a recycler is exported to a non-OECD country such as China. (Basel Secretariat 2010, p. 5 of response by BAN)

It is worth noting that it was BAN itself that drew attention to the conditions of production of this statistic because the figure continues to circulate inside and outside the e-waste literature as if it were a well-established fact. Yet, as BAN's own commentary points out, the 50–80 percent of exports figure was derived from highly circumscribed circumstances (it came from a single interview with an industry insider) and refers to a very different geography of export trade than that ascribed to Xiezhi and co-workers in the draft guidelines. Indeed, BAN's comment implies that Xiezhi and co-workers mischaracterize the geography of exports to which the 50–80 percent figure refers (in the BAN report it refers to exports from the "western U.S."; see Basel Action Network 2002, 1). Xiezhi and co-workers (2008) themselves cite two sources for the 50–80 percent figure: BAN's (2002) *Exporting Harm* report and a UN pamphlet titled *E-Waste, the Hidden Side of IT Equipment's Manufacturing and Use* (Schwarzer et al. 2005). Schwarzer and co-workers (2005), like Xiezhi and co-workers (2008), cite BAN's 2002 report among their sources for claims about the scale of the export of e-waste (though they do not refer to the 50–80 percent figure). There is evidence here of a kind of circular reference chain that has had real consequences for how knowledge about the e-waste trade has been shaped, an issue I address in more detail in chapter 5.

The grounds of the 50–80 percent figure were disputed by Canada, by PC Rebuilders & Recyclers (PCRR, a Chicago-based computer repair and recycling business), and by the United States (see figure 2.10 online). Canada argued that "while it is not disputed that export does occur to developing countries and countries with economies in transition, it is questionable whether these figures [50–80 percent of exports] are reliable given the complexity of international trade and increasing international efforts to combat illegal flows of e-waste" (Basel Secretariat 2011b, p. 2 of response by Canada). PCRR was more emphatic, commenting, "These numbers [50–80 percent exports] are pure guesses" (Basel Secretariat 2011b, p. 1 of PCRR's response).

Despite the comments from BAN and the counterpositions expressed by PCRR and Canada, the next draft of the guidelines not only kept the 50–80 percent figure but rewrote the introductory text in two important ways. First, the introduction was redrafted to add a direct reference to BAN's *Exporting Harm* as the source for the 50–80 percent figure. Second, Xiezhi and co-workers' study was recharacterized as a confirmatory source of that figure. The May 8, 2012, draft of the guidelines read in part that Xiezhi and co-workers "confirm these practices [referring to the 50–80 percent figure from BAN] and suggest similar percentages of export" (Basel Secretariat 2012a, 3). The text of this new draft of the guidelines implied that Xiezhi and co-workers had performed a study similar to the BAN study and confirmed the 50–80 percent of exports statistic. Yet they had done no such thing. They did not investigate the export trade of e-waste at all. The purpose of their study was quite different. Instead of the export trade, Xiezhi and co-workers' study concerns soil contamination around sites of open burning of e-waste materials in Guiyu, China (a site highlighted in BAN's *Exporting Harm*). Rather than investigating the export trade, Xiezhi and co-workers cited the 50–80 percent figure (and BAN's *Exporting Harm*) in the opening sentence of the paper as part of the justification for their study of soil contamination. So Xiezhi and co-workers confirmed the 50–80 percent figure only to the extent that they cited BAN's report, did not question its veracity, and then went on to study soil contamination but not trade. Under these circumstances, the United States strongly opposed this change to the guidelines' introductory text. The United States wrote emphatically, "We support the deletion of this paragraph [referring to BAN/Xiezhi and co-workers' figures]. The statistics used in this paragraph have been continually cited by the media and NGOs; however, there is no qualitative analysis to back them up" (Basel Secretariat 2012a, p. 5 of the U.S. response; see figure 2.10 online).

The United States' characterization of the 50–80 percent figure is not hyperbole. As I discuss in detail in chapter 5, that figure has traveled very widely despite the qualifications added by BAN during the internal discussions over the guidelines cited above. The report *Exporting Harm*, in which the figure first appeared, is widely cited in the academic literature on e-waste, for example, where it consistently ranks among the top ten cited references despite not being a peer-reviewed article. The figure has also been treated as an established fact in national policy debates in the United States (U.S. Congress 2009, 5, 12). Both *Exporting Harm* and the 50–80 percent of exports figure operate as a ground floor for characterizing the problem of e-waste in international regulatory debates in the negotiations over the technical guidelines. Since my interest for the moment is to continue diving in the magma of actors forming alliances and oppositions around e-waste as a matter of concern, as evidenced in the guidelines, I return to the broader import of *Exporting Harm* in chapter 5.

Used Equipment Becomes a Cosmopolitical Issue

A cosmos is a vision of the world as a well-ordered whole. The positions of actors in the debates over the technical guidelines articulate different cosmoses. Not all of those visions are mutually exclusive, but some are. Where cosmoses conflict on mutually exclusive grounds, struggles over whose worlding(s) will prevail become prominent. Arguably, the specific issue of used electronic equipment became this kind of cosmopolitical matter during negotiations over the guidelines.

Several groups of arguments emerged in the first draft of the technical guidelines around the issue of how used equipment should be handled. These groups of arguments included whether parties to the Basel Convention can independently or voluntarily define waste, the need for functionality testing of equipment before export, and how export for maintenance, refurbishment, or repair is to be managed (see figure 2.11 online). The first draft of the technical guidelines included a provision for a voluntary notification procedure relating to exports for "repair, refurbishment, or upgrading" (Basel Secretariat 2010, 9). The provision was opposed by BAN and ITI, but for quite different reasons. BAN was adamant: "There is no legal basis for the so-called voluntary procedure. ... It is not up to the Parties to independently decide what is a waste under the Convention and what is not" (Basel Secretariat 2010, p. 10 of response by BAN). BAN noted that the Basel Convention only "allows countries to independently augment the baseline definitions" of the convention (Basel Secretariat 2010, p. 10 of response by BAN). ITI expressed a rather different concern: that equipment suitable for repair and reuse would be designated as waste and thus be channeled to recycling. ITI argued that if the Basel Convention's definitions of waste were expanded to include equipment currently managed through repair and reuse, "the practical effect will be to limit the useful life of products leading to premature and accelerated waste generation" (Basel Secretariat 2010, p. 11 of response by BAN).

A second, related set of arguments emerged around the need for functionality tests to distinguish waste from nonwaste electronics. BAN held that any equipment not tested or that fails a full functionality test should be considered waste. Several actors were concerned, however, that full functionality was too strict a guideline and would result in the premature destruction of equipment for recycling that was otherwise suitable for reuse. ITI argued that a requirement of full functionality "could have the unintended effect of directing functional equipment requiring only minor repair or refurbishment to materials recovery before the end of its useful life, with unnecessary environmental costs and lost social, economic and environmental benefits" (Basel Secretariat 2010, p. 9 of response by ITI). Canada held that the meaning of "fully functional" needed to be clarified because equipment is often multifunctional; for example, "camera phones may still function well as either a phone or camera even if one of these capabilities does not work" (Basel Secretariat 2010, p. 2 of response by Canada). Both Norway and the United States advocated for what they called a key functionality approach. The United States used the same example of a camera phone as Canada to make the point that even without a working camera, the phone would be suitable for reuse as a phone (no guidance was provided, however, about what constitutes a key function—is a camera phone with a working camera but that cannot make calls considered waste or not waste?).

The most serious dissent against the criteria of full functionality was brought by the American Chamber of Commerce to the European Union (AMCHAM EU). The group argued that passing a full functionality test prior to export "would be an impossible requirement to meet" and, were it to become part of the guidelines, it "would effectively shut down the repair and refurbishment market" (Basel Secretariat 2011b, p. 3 of response by AMCHAM EU). One might interpret such a position as a kind of cynical defense of industrial interests, but the situation is more complex than that. Substantial repair and refurbishment markets exist for equipment that is under warranty, leasing, or product servicing agreements. This is the case for consumer electronics, and it is also the case for medical devices. AMCHAM EU further pointed out that manufacturers of medical devices were legally required to investigate "adverse events" in which patients may have been harmed by the functioning (or malfunctioning) of a medical device. The infrastructure for investigating adverse events is unevenly distributed globally. That is, because of the highly specialized character of medical devices, testing, repair, and refurbishment facilities exist in only a few places rather than in every country in which a device manufacturer sells such equipment (broadly speaking, the same is true for consumer electronics). As Philips, a major player in the medical device field, argued:

The proposed guidelines will seriously threaten the movement of used medical devices destined for refurbishment and remanufacturing. ...

Used medical devices are refurbished using the highest possible international standards and sold under full warranty equal to new. Philips refurbishing program provides reliable and cost effective refurbished medical devices, allowing more patient access to up-to-date technology. This program relies on transboundary movement of used professional equipment to Philips' refurbishing locations. Defining used professional electronic equipment destined for refurbishing or repairs as "e-waste" under the proposed guidance threatens to disrupt and may stop legitimate transboundary movement of this equipment, prematurely diverting valuable equipment to waste recycling channels. Medical device refurbishment and repair is an effective means of reducing e-waste while ensuring greater global access to medical device technology. If adopted, we believe this approach would impose new and unjustified barriers to legitimate international trade in used equipment without providing any significant benefit. (Basel Secretariat 2011b, p. 1 of response by Philips)

Philips would later claim that interrupting the transboundary movements of medical devices for repair, refurbishment, and reuse would increase health care costs by €500 million annually (Basel Secretariat 2012b, p. 3 of response by Philips. Philips provided no evidence of how this figure was reached). The Global Diagnostic Imaging, Healthcare IT and Radiation Therapy Trade Association (DITTA) made a similar claim, arguing that the repair, refurbishment, and reuse of medical devices reduced health care costs by 20–30 percent (Basel Secretariat 2015a, p. 1 of response by DITTA. Again, however, no indication is given as to how this figure was reached).

Representatives of the medical device field argued that a solution could be found by exempting particular classes of devices from the technical guidelines, including medical equipment, "professional" equipment, and equipment under warranty, leasing, or product servicing agreements (see figure 2.12 online). BAN, in contrast, strongly opposed exemptions of any kind. Of fundamental concern to BAN was that allowing exemptions into the technical guidelines would set a precedent that would undermine the Basel Convention itself. The group claimed that "since proposing this exception, we have seen industry trying very hard to propose other loopholes and expand this exception to the point where very large amounts of hazardous waste electronics could be exported. This is unacceptable" (Basel Secretariat 2011b, p. 9 of response by BAN). BAN pointed out a number of issues the group saw in the exemptions being proposed that created the possibility of loopholes, such as the problem of imprecise definitions, noting, "Most IT equipment on earth today can be considered equipment for 'professional use' and thus the exemption is open-ended" (Basel Secretariat 2012a, p. 3 of response by BAN). Ultimately, BAN argued, "exports either need to be shown to be fully functional or non-hazardous to fall outside of Basel Convention rules" (Basel Secretariat 2012a, p. 3 of response by BAN).

The outlines of mutually exclusive cosmoses can be discerned in this disagreement over whether or not exemptions should be permitted. As I discuss in more detail below, one such cosmos cleaves to the Basel Convention's statist geography of signatory parties, dividing the world into Annex VII and non–Annex VII states. Another cosmos envisions a world organized by logistics networks. These networks bring digital equipment to markets and return that equipment to a set of sites unevenly distributed around the world for specialized repair, reuse, and refurbishment work. A list of seven criteria to tightly define the conditions under which used equipment, particularly medical devices, could be exempt from the Basel Convention was proposed in an effort to find common ground among these competing visions of how the world ought to be. These criteria appeared in paragraph 26b of the draft guidelines of November 26, 2014 (Basel Secretariat 2014, 10–11). In their comments, actors refer to this paragraph using the shorthand of the "26b preferred option" (see figure 2.13 online).

In brief, the criteria in 26b specify guidelines relating to the following:

1. *Notification:* Parties must notify the Secretariat of the Basel Convention that they do not consider used equipment to be waste.

2. *Compliance:* Equipment to be exported must comply with the Restriction of the Use of Certain Hazardous Substances (RoHS) legislation.

3. *Ownership:* The exporter remains responsible for all exported equipment and components, as well as for waste arising from testing, repair, and refurbishment processes until equipment was tested to be fully functional or made available for direct reuse.

4. *Contract:* Each export shipment must involve a valid contract between exporter and the importing facility.

5. *Declaration:* Each export shipment must include a written and signed statement by the exporter guaranteeing all provisions of paragraph 26b are fulfilled.

6. *Disposal:* All waste arising from testing, repair, or refurbishment activities must be dealt with in an environmentally sound manner.

7. *Packaging:* Each piece of equipment and its components must be packaged so as to protect it from exposure to hazards or damage to equipment.

These seven criteria, listed in paragraph 26b, enjoyed support from BAN, DigitalEurope, ITI, Norway, and in general all other actors as well, though a number of disagreements were registered over specific criteria or parts thereof. The "preferred option" represented an emerging common world, that is, a cosmos that might have been shared by the otherwise disputatious actors. Ultimately, however, this cosmopolitical compromise did not hold, and the negotiations over the guidelines collapsed at COP 12.

Discussions around the first criterion (notification) included a proposal to maintain a list of all facilities permitted to receive used equipment. The EU, the United States, and Japan argued that maintaining such a list would not be feasible because the number of facilities would be very large and the list would require constant updating as facilities' permissions changed. The EU and Japan objected to the requirement of the second criterion (compliance) that used equipment be RoHS compliant because of how this would negatively affect transboundary movements of used medical devices, particularly those containing CRTs. The demand for compliance with RoHS, which is part of EU legislation, was also deemed to introduce a problematic Eurocentric orientation to the guidelines by Argentina, Brazil, Japan, and the United States. Meanwhile, Korea opposed the requirement under the third criterion (ownership) that all unusable equipment be treated as waste. Against parts of the sixth criterion (disposal), Korea argued that "it is premature to add a condition that all residual waste shall be disposed of in an Annex VII country, when the Annex VII has not entered into force yet" (Basel Secretariat 2014, p. 1 of response by Korea). Similarly, the United States stated that while "we strongly support the idea that all waste should be managed in an environmentally sound manner … we do not support a reference to the Ban Amendment in the language as it is not in force" (Basel Secretariat 2014, p. 4 of U.S. response).

Despite these disagreements over some specifics of paragraph 26b, no actor disagreed with its overall attempt to define criteria for exempting used equipment from the Basel Convention. BAN, however, articulated a maximalist position in support of the seven criteria. The organization argued that the criteria constituted a package, not a list of individual measures (Basel Secretariat 2014, p. 2 of response by BAN). The group maintained this position because "exercising exemptions as is being proposed in 26b is to exercise derogation from the norm [of the Basel Convention]" (Basel Secretariat 2014, p. 2 of response by BAN). BAN argued that the seven criteria needed to be understood as a package to stave off what it saw as a dangerous trajectory potentially being built into the guidelines through the provision for exemptions. The group was deeply concerned about the danger of "giving up jurisdiction and control over one fraction of the most traded and potentially dangerous waste stream [e-waste] subject to transboundary movement today [because] declaring something a non-waste means that it falls outside of the Convention and all of its obligations" (Basel Secretariat 2014, p. 2 of response by BAN).

Important changes in the draft text of the technical guidelines occurred between the release of the November 20, 2014, version and the version under consideration at COP 12 in May 2015. Exactly how such substantial changes came about, however, is not captured in the record of meeting documents. A small intersessional working group met in Konstanz, Germany, in January 2015 specifically to address paragraph 26b of the draft technical guidelines (Basel Secretariat 2015c, 2). According to the record of that meeting, discussions "indicated there is wide support for some elements of what is presented as the 'preferred option' [paragraph 26b] ... while on other elements there are still differing views" (Basel Secretariat 2015c, 3). No mention is made of any major changes to the text. Yet between the Konstanz meeting and the draft technical guidelines circulated at COP 12, major changes appeared in the text of the guidelines precisely on the seven criteria previously addressed in paragraph 26b.

The draft circulated and eventually adopted at COP 12 both reduced the number of criteria determining used equipment as nonwaste and substantially altered the specifications of those criteria it did keep or introduced. Instead of the seven criteria originally articulated in paragraph 26b, this draft listed a total of six criteria, four relating to export for direct reuse or extended use by the original owner and two criteria relating to export for failure analysis, repair, or refurbishment. In brief, the criteria included the following:

1. A copy of the invoice and contract must be available prior to and during transport.

2. Evidence of testing.

3. A declaration by the individual responsible for shipment that its contents are not considered waste in any countries involved in the shipment.

4. Appropriate protection (i.e., packaging) to avoid damage of equipment and exposure to hazards.

5. Appropriate documentation prescribed elsewhere in the guidelines.

6. A valid contract between exporter and receiving facility.

Consensus could not be achieved around these revamped criteria for adjudicating the distinction between waste and nonwaste used electronics at the COP 12 meeting, and the process collapsed. As a direct result, the guidelines were adopted on an interim and non–legally binding basis.

The shattering of the world proposed in the "preferred option" brings into sharp relief the range of cosmoses being articulated by the actors' positions and the degree to which given actors expressed support (i.e., partisanship) for given worldings. Two cosmoses are discernible from all the actors' positions, one I dub "Mono-Bloc World" (MB World) and one I dub "Global Value Chain World" (GVC World; see figure 2.14). Both are utopias in that neither actually exists beyond the stated positions of the actors themselves. Nor do these two cosmoses exist at the opposite ends of a smooth continuum. The range of differences between them is magmatic, pouring smoothly into shape here, bursting suddenly into new shapes there; melting tunnels through foundations laid earlier; erupting unexpectedly through layers of previous consensus.

MB World is the world of the Basel Convention and the Ban Amendment. Like the guidelines themselves, the Ban Amendment has been adopted by the COP but not yet ratified by enough signatories to bring it into legal force. In that sense, MB World is a utopic cosmos in that it exists partially rather than in full. As described earlier, Annex VII divides signatories into two monolithic blocs of states. The signatories to Annex VII are the member states of the European Community, the OECD, and Liechtenstein. Non–Annex VII states are all other signatories to the Basel Convention. The Ban Amendment prohibits trade from Annex VII to non–Annex VII signatories. The amendment is also entwined with principles for the environmentally sound management of waste articulated elsewhere in the convention. Those principles include self-sufficiency, proximity, and least transboundary movement. Together these principles articulate an autarkic approach to managing waste subject to the Basel Convention—that is, signatories should handle waste within their own borders. Thus, partisans of MB World can be recognized by the extent they can be imagined shouting (or whispering) support for the slogan, "Autarky!"

GVC World is a cosmos of transboundary trade in used electronic equipment. It, too, only partially exists. Partisans of this worlding argue that substantial industrial traffic in used electronics already exists, for example, to allow root cause analysis in the medical field or for repair and refurbishment of devices under warranty. Those activities underpin and rely on industrial distributions of testing and repair facilities that do not match up to the geographies of MB World. GVC World is a cosmos of reverse supply chains rather than statist geographies of the Basel signatories. Its partisans are recognized by the extent they proclaim, "Reverse supply chains!"

Ultimately, it was an inability to gain consensus over how the technical guidelines would distinguish waste from nonwaste used electronic equipment that lead to the breakdown of negotiations at COP 12. The competing cosmoses of the actors literally could not coexist. One worlding attempts to instantiate a geography of transboundary trade consisting of a network of cross-border movements for repair, refurbishment, or reuse (see figure 2.14). Another worlding attempts to actualize a different geography of transboundary movements. It is a binary world in which two monolithic blocs of sovereign territories, Annex VII and non–Annex VII, may trade intraregionally but not interregionally (see figure 2.14). These different worldings represent the different cosmoses of the actors and are fundamentally incompatible with one another. Actors favoring repair, refurbishment, and reuse argued that their entrenched business models depend on a geography of transboundary movements for repair, refurbishing, and reuse that does not match up with the Basel Convention's prohibition of transboundary shipments of equipment deemed to be waste. Consequently, unless rules governing equipment destined for repair, refurbishment, or reuse were carefully crafted those rules would interfere with what these actors understood as legitimate commercial trade. At stake for them is a particular worlding or cosmos of used electronic equipment, a cosmos fundamentally threatened by rules that treat used equipment as waste.

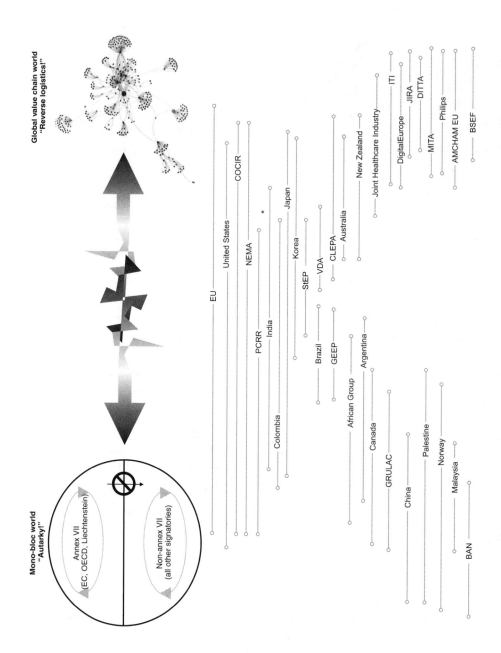

◀ **Figure 2.14**
Diagram providing a qualitative representation of actors' partisanship for a preferred worlding. Horizontal lines extending from each actor's name indicate that actor's apparent willingness to negotiate over the two options. The position of a given actor's name within its horizontal line (e.g., EU) indicates its relative preference within the overall range it appears willing to negotiate. The two options (Mono-Bloc World and Global Value Chain World) do not sit at opposite ends of a smooth continuum. Instead, the jaggedness of the arrow between them is intended to visually represent that the two options are separated by a patchy, distributed, and noncoherent set of options. The lower down horizontally actors are positioned, the less they share of a common worlding.

Conclusion

The stakes of distinguishing waste from nonwaste electronics at COP 12 were not mere quibbles over technical *wordings* but instead fundamental disagreements over *worldings*. Two years after COP 12 the parties met again between April 24 and May 5, 2017, as part of a joint meeting of the Basel, Rotterdam, and Stockholm Conventions. In the final decision made at this most recent meeting, the COP agreed to "establish an expert group to look further into the TGs [Technical Guidelines] to advance the work towards finalization of the guidelines" (IISD Reporting Services 2017, 20). Who will make up this group of experts? Will they represent the same interests as those of the actors charted in this chapter? How much and what kind of work will lead to the goal of finally distinguishing between waste and nonwaste electronics? As I write, all of these remain open questions. The controversy, in other words, remains ongoing, and still no legally enforceable distinction between waste and nonwaste used electronics exists in the Basel Convention. The partisanship of the actors, however, is likely to be poorly judged by whether or not they have an economic interest in a given outcome. Mapping the controversy over the technical guidelines as it unfolded between 2010 and 2015 showed, among other things, that all actors had some degree of economic interest in play.

Several developments seem likely to follow from the most recent COP decision to reconvene a group of experts to finalize the technical guidelines. First, the inherent indeterminacy of hazardousness and of waste will continue to complicate future negotiations over the distinction between waste and nonwaste. Second, mapping the controversy surrounding the negotiations over this distinction suggests that disparate actors—be they corporations, trade associations, NGOs, or states—remain convinced that an underlying problem is a lack of precision of definitions. The public record of documents submitted as comments on various drafts of the technical guidelines do not suggest that the call for more precision can simply be dismissed as a shrewd or cynical strategy of any one actor or group of actors to stave off consensus and thus to forestall implementation of these guidelines. Instead, it seems, actors widely share the view that if only the meanings of "hazardous," "waste," and "nonwaste" could be articulated with higher degrees of exactness, disagreements at the heart of the controversy would be resolved and a common worlding shared by all actors could be found. Wynne's (1987) analysis of hazardous waste regulation suggests otherwise. The search for precision will more than likely generate a "cycle of self-destruction of institutional credibility" (Wynne 1987, 12). This seems a real risk with respect to the historical and future trajectory of the technical guidelines on e-waste at the Basel Convention. The search for precision at the point of export defers action that could be instituted much further upstream, before that which will become waste is made in the first place. International agreements that influence production processes—such as prohibiting the use of particular chemicals or taxing wastes arising in production at high enough rates to influence design and manufacturing—are needed. Here I am merely gesturing to possible solutions to "the e-waste problem." How that problem is formulated as such plays a dramatic role in how solutions to it are, or can be, imagined. The inherent indeterminacy of e-waste suggests it is necessary to defamiliarize the taken-for-granted notion that it is strictly a problem of postconsumption disposal.

Yet, while defamiliarization of e-waste is necessary, it is not an end in itself. My point in raising questions about precision and corollary concepts such as certainty over the waste/nonwaste distinction is not about fomenting doubt that might feed its cynical use to forestall decision making about solutions to e-waste. My point is quite the opposite. Because of the fundamental indeterminacy of waste, doubts of some kind cannot be avoided. This is a situation invoked by the tonnage, toxicity, heterogeneity, and harms that characterize not just e-waste problems but contemporary waste problems more generally. The question now becomes, what is the right thing to do with e-waste *despite* such fundamental indeterminacy?

3 The Discard Test

If two objects that go under the same name clash, in practice one of them will be privileged over the other.
—Annemarie Mol (2002, 47)

And if the global were small and noncoherent?
—John Law (2004b, 13)

Introduction

On June 19, 2014, Joseph Benson, a Nigerian citizen running an electronics recycling and repair operation in the UK, was sentenced in the Crown Court at Snaresbrook, London, to a fine of £142,145 (about U.S. $240,000) and jail time of sixteen months for attempting to export eleven shipping containers of discarded electronics to various destinations in West Africa. His conviction stemmed from his business becoming caught up in a media event staged by Greenpeace in collaboration with two UK news outlets and in an investigation by the UK Environment Agency code-named Operation Boron (UK Environment Agency 2012). Greenpeace, the *Independent*, and Sky News had an engineer render an old television nonworking and insert a tracking device into it. The television was then left at a civic amenity site in the UK. It was subsequently tracked to Alaba market, a bustling cluster of electronics repair and refurbishment shops in Lagos, Nigeria (Mansfield 2012; Milmo 2009b). The media event entered the court record of Benson's trial when a Sky News engineer testified that the television in question "could not be economically repaired in the UK" (Lord Justice Pitchford 2012, 7 para. 17).

The presiding judge at Benson's sentencing hearing noted:

I quite see that these crimes are commercial crimes to a certain extent in that no sort of measurable individual is hurt by the crimes that are committed, but on another level these crimes are quite serious for society as a whole because our children and grandchildren are going to be affected by environmental damage across the world and we are affected by environmental damage. (Dawson 2014, 4, 10)

Judge Dawson's words turn the case into a parable of global intergenerational justice. After Benson's conviction, the UK Environment Agency issued a press release that was picked up by the *Guardian* and later highlighted in a major report on global waste crime put out by the United Nations. An Environment Agency official quoted in the press stated that Benson's "sentence is a landmark ruling because it's the first time anyone has been sent to prison for illegal waste exports" (*Guardian* 2014; UK Environment Agency 2014, 1; United Nations Environmental Programme 2015, 42). How does a jumble of unlike things—a deliberately disabled television, a Nigerian citizen and businessman, an export business, and a UK regulatory authority, among other actors—get gathered together into a narrative that coheres as such a global story? A partial answer lies in how the case worked out its jurisdiction over the distinction between two objects, waste and nonwaste, that fall under the same rubric, discarded electronics.

The allegorical character of the case leads me to explore several interlinked questions in this chapter. First, how do the case and the legislation relevant to it approach and deal with electronic waste? Second, insofar as the law, broadly speaking, pursues "goods" and fights "bads," what goods are sought and what bads are fought in the ways the case approaches waste legally? Third, how are these goods and bads set up as such—that is, what are the grounds on which they are based? Finally, what are the implications of the goods pursued and the bads repudiated in the case? These questions are queries that drive investigations of what science and technology studies (STS) scholar Annemarie Mol (2002) describes as the "politics-of-what." By this she means to draw attention to how the practical work of framing a problem enacts or brings into existence the objects of the problem so framed. Such ontological politics, to use Mol's (2002) term, are crucial because particular framings make some proposals for solutions thinkable and, at least potentially, actionable. At the same time, specific framings can foreclose other possible framings and the solutions they might suggest. Alternatives, in other words, might not be just left unexplored; they may even become unimaginable. By tracking the practical work of enacting objects into being, Mol is interested in keeping the making of the actual in view. She is also keen to highlight how the institution of the actual is a practical achievement, and therefore, how actual objects are riddled with provisionality and contingency. The actual is an achievement rather than a given. If it can be made, then it could also be made otherwise—even if not necessarily easily so (Sismondo 2017).

The practical action of making the actual leads me to what the legal scholar Mariana Valverde (2009) calls the work of jurisdiction. The term refers to the action that organizes jurisdiction, legally speaking, into a well-ordered whole, or what anthropologists sometimes call a cosmology (Douglas 1966; Ferguson 1999) and what I referred to in the previous chapter as a cosmos or worlding. In the work of jurisdiction, worldings of e-waste are at stake. Valverde advocates following the action that coordinates jurisdiction in ways relevant to a given legal case. In this sense, the work of jurisdiction can be thought of as a script for coordinating legal action, or, said differently, for giving it coherence (see also Akrich 1992; Latour 2012). The script is composed of questions about the *what*, the *who*, and the *where* of a legal dispute. These questions work as sorting and order-ing devices for organizing the practical work of acting legally. They can be posed in any order, but once any of them is answered, questions about *how* to coordinate jurisdiction into a well-ordered whole for the case at hand get resolved (Valverde 2009). Even if there may be many other ways of resolving this coordination in principle, one way of organizing jurisdic-tion will prevail over those other possibilities.

The public record of Benson's case shows it to be entangled with four other jurisdictions that distinguish waste from nonwaste: that of the Euro-pean Union, the Organisation for Economic Co-operation and Develop-ment (OECD), the Basel Convention, and the Bamako Convention (a treaty among African nations undertaken largely as a response to perceived weak-nesses in the Basel Convention; see Clapp 1994). These entanglements are important. The other jurisdictions with which Benson's case is entangled do not all approach and deal with waste and nonwaste the same way. The imbroglios of the case make it helpful to interpret it as an attachment site, a situated locus of entanglements (Haraway 2008, 287; 2010, 53). If we stick with the trouble of entanglement, as Haraway advises, we can assess the practical work the judges and juries must do in Benson's case to dis-entangle the convolutions of the case into a well-ordered whole in which a UK-based legal case stands in for global justice. At the same time, sticking with that trouble of the case also demonstrates that clash and noncoherence are only provisionally and contingently held at bay, even as they are folded together into the legal technicalities of jurisdiction over the distinction between waste and nonwaste.

The Discard Test

On September 25, 2012, Benson's earlier indictment by jury at Basildon Crown Court under Judge Jonathan Black was heard on appeal. Benson and his codefendants had been indicted under Regulation 23 of the UK's Transfrontier Shipment of Waste Regulations (2007) (hereafter the UK Transfrontier Act). Though lawyers for both the prosecution and the defense agreed that the appellants were handling equipment deemed waste under the law, Benson and his colleagues had been buying that equipment from civic amenity sites where consumers could drop off, free of charge, unwanted items (including electronics) for recycling and disposal. According to one media report, Benson's company, BJ Electronics, had been paying such sites between £1 and £5 for computer monitors, televisions, and stereos (Milmo 2009a). Those rates match what the Environmental Investigation Agency, a London-based ENGO, found to be the "actual costs of recycling a CRT" legitimately in the UK (Environmental Investigation Agency 2011, 11). So the usual way of explaining e-waste exports, as a means of avoiding costs for environmental compliance, do not seem to have been the issue in Benson's case. Consumers were dropping off items free of charge and Benson's business was paying the same rates for such items as the actual costs of recycling them in the UK. What was at issue was the distinction between waste and nonwaste under the relevant law and how locations at which the equipment was gathered, along with the destinations to which that equipment was being shipped, played a role in that distinction.

To define waste, the UK legislation refers to Article 36(1) of the European Waste Shipment Regulation 1013/2006 (hereafter EC 1013/2006) which itself traces a definition of waste to two earlier pieces of EU legislation, the Waste Framework Directive of 1975 and its 1991 amendment. In the EU legislation, waste is defined as "any substance or object ... which the holder discards or intends or is required to discard" (European Union 1991,

2). As Cheyne and Purdue (1995) show, the 1991 amendment entailed a key shift in the legal definition of waste in the EU and its member states. The amendment changed the operative word from "dispose" to "discard." On the one hand, the shift to "discard" was seen to offer a wider umbrella than did "dispose" under which EU member states could enact nationally specific legislation. On the other hand, "no substantive link" was established in the directive between the definition of waste and the definition of disposal operations (Cheyne and Purdue 1995, 158; see also Fluck 1994).

A 1997 ruling by the European Court brought legal—if not semantic or philosophical—clarity to the question of the relationship between the actions of disposal, discard, and the meaning of waste within the EU. Advocate General Jacobs concluded that the term "waste" in the EU Waste Framework Directive applied to

any substances or objects which the holder discards or intends or is required to discard, even where they are capable of re-use and may be the subject of a legal transaction or quoted as being of commercial value on public or private commercial lists. …

[Waste] is not to be understood as excluding substances and objects which are capable of economic reutilization. A residual substance derived from a production or consumption cycle in a manufacturing or combustion process constitutes "waste" and is subject to the system established by the [European] Community rules if its holder discards it or intends or is required to discard it. (Advocate General Jacobs 1997, 3583–3584)

Jacobs's ruling established the discard test as the key grounds for distinguishing waste from nonwaste in the EU. The discard test would be of central importance to the work of jurisdiction at Benson's trials.

There were three grounds on which Benson's defense sought to appeal his earlier conviction: the waste versus nonwaste distinction, the meaning of recovery, and the question of strict versus corporate liability (i.e., whether Benson personally rather than his business should be found liable). It was the discard test that formed the basis of the judges' ruling in Benson's original trial and his appeal. The written decision for Benson's appeal trial is twenty-seven pages long. Ten of those pages are devoted solely to answering the question of the ontological status of the equipment found in the containers being shipped by Benson and his colleagues. Was the equipment waste or nonwaste?

The judges in Benson's appeal case cited an earlier appeals case (Lord Justice Moses 2008) as precedent for the definition of substances as waste even if they can be economically reused or have commercial value. In that case, the UK Environment Agency appealed a decision against it and in favor of Thorn International, a used electrical goods business. The Environment Agency had charged Thorn International with violations of the UK Environmental Protection Act and the EU Waste Framework Directive for handling waste without a waste management license.

Thorn International's business model was to buy used electrical items it deemed to be functioning or capable of repair from another business called Wincanton. Wincanton in turn derived its stock of electrical items from retail returns, that is, through contracts with electrical appliance retailers that had supplied replacement products to customers. Wincanton would take the returned items to a central warehouse and test them for repairability. Items deemed to be beyond repair were disposed of through licensed waste carriers while those able to be repaired were sold to "the refurbishment and resale trade," including Thorn International (no mention is made in the case of transboundary shipment, so it appears that Wincanton and Thorn International were engaged only in the domestic UK refurbishment and resale trade; see Lord Justice Moses 2008, para. 9).

Once at Thorn International, electrical equipment would be stored on-site for up to, but no more than, one week while it awaited repair for resale to the public. While Thorn never stored small items such as computers outside, other electrical items such as refrigerators might be stored in an outdoor yard for up to seven days awaiting repair. The latter situation of outdoor storage is what drew the attention of Environment Agency officials and led them to demand of Thorn that the company be in possession of a waste disposal license. The agency then made two claims against Thorn's operations. The first was that "the electrical items had been discarded by the consumer (the householder) at the moment he made a contract with the retailer to exchange his particular refrigerator, for example with a new one from the retailer" (Lord Justice Moses 2008, para. 15). In the eyes of the Environment Agency, retail return constituted discard by the householder, and discards are waste under the relevant UK and EU law. In other words, the electrical items being repaired and resold by Thorn had become waste at the retailer the moment a consumer returned an electrical device to a given retailer. The second claim of the Environment Agency was that Thorn's selection and purchase of items it deemed repairable were not sufficient activities to transform the equipment from waste to non-waste. The Environment Agency contended that the items "would only cease to be waste at such a time as the trained engineers (at Thorn) had completed such repair or refurbishment as was necessary" (Lord Justice Moses 2008, para. 15).

Let's consider what the Environment Agency's interpretation of the law suggests. It would seem that had Thorn stored all equipment inside, the equipment would not have come to the attention of Environment Agency officials and then deemed to be waste; it also seems that to avoid a repair business being deemed a waste handling operation by the Environment Agency, the activity of repair or refurbishment must be completed instantaneously at the moment a householder undertakes action that constitutes discard. An implication of this doctrine of discard is that in the case of electronic equipment, if its holder can be said to have discarded it, it is waste, regardless of its characteristics as working, repairable, or reusable equipment. A day-old flat screen with a scuff mark on it discarded by its holder is waste, just as a thirty-year-old CRT monitor with a shattered screen discarded by its holder is waste; a working smart phone with a cracked but repairable screen is waste if it is discarded, just as is an unsorted box of mismatched copper wiring. The meaning of waste is not in the stuff itself. It derives only from what its holder has done with it, irrespective of whether the equipment might be economically reusable.

Judges Moses and Blake agreed with the Environment Agency up to a point. They noted that "the question of whether something which is undoubtedly waste ceases to be waste is determined by whether the cycle of repair or restoration is complete" (Lord Justice Moses 2008, para. 21). But, Judge Moses continued,

I reject the contention that the justices were bound to conclude that the electrical goods were waste in the circumstance that there was a contractual arrangement between consumer and retailer, who agreed to take the particular item in question on purchase of another new item. ... The justices found that these items had not been discarded. Whether they were focusing their attention on the moment at which they had been exchanged with a new item obtained from a retailer or at a later stage is not clear. But for my part I would reject any principle which established that the justices were bound to conclude that at that stage they had been discarded. (Lord Justice Moses 2008, para. 22)

Ultimately, Judges Moses and Blake ruled against the Environment Agency and in favor of Thorn. They came to their judgment because they agreed with the original court's decision that the action of product return by householders to retailers does not necessarily constitute discard. Consequently Thorn, the judges ruled, was not in need of a waste handling license because the items Thorn and its supplier, Wincanton, handled had never become waste.

The Thorn case might seem to have had a logical connection to Benson's own and to act in his favor. However, the judge in Benson's case ruled that Thorn differed from Benson's own situation. In Benson's case, he and his colleagues had been collecting electrical items at civic amenity sites rather than from returns to retailers. The action of depositing materials or equipment at civic amenity sites constitutes discarding within the meaning of UK law—and that meaning also has the effect of transforming the materials and equipment placed at these sites into "waste" under that law. Crucially, neither the prosecution nor Benson's defense attorney disputed this point. Both sides agreed that the items Benson and colleagues collected had been discarded by their previous holders and were, as a consequence, "waste" (Lord Justice Pitchford 2012, para. 32). At Benson's appeal trial a key question then became whether what Benson and colleagues "had done to those objects before loading rendered them non-waste" (Lord Justice Pitchford 2012, para. 51). In other words, had Benson or his colleagues done some practical work that changed the ontological status of the objects in question from waste to nonwaste? We know the short answer to this question since Benson and his colleagues lost their appeal and were indicted on waste trading charges, and Benson was sentenced to jail time. But how did the court come to the conclusion that Benson and colleagues were shipping waste, rather than this curious category called "nonwaste"?

At Benson's appeal trial in June, both the lawyers for the prosecution and those for the defendants were in agreement that Benson and colleagues had collected waste because the original holders had discarded the equipment at civic amenity sites. Under the law, the condition of the equipment was irrelevant. Even if the equipment was working, reusable, or repairable, it was for the purposes of law made waste by the act of discard. What was in dispute was whether the appellants had performed activities sufficient to shift the collected equipment from the waste to the nonwaste category. The defense argued that by visual inspection and selection at the civic amenity sites and by subsequent testing, the appellants had performed activities sufficient to shift the equipment to the nonwaste category. The prosecution disagreed. It argued that EU guidelines for shipping e-waste necessitate proper testing for key functions prior to export (see European Union 2007, Article 11-16). Equipment that fails such tests remains "waste" until some action such as repair is taken to shift it into the nonwaste category. Ultimately, the prosecution argued that "from the condition of the goods on inspection of the contents of containers in which they were consigned for shipping to Nigeria, that they remained waste in the hands of the collectors and shippers (Benson and colleagues)" (Lord Justice Pitchford 2012, para. 32). About half the equipment items in the containers in Benson's case were found to fail tests of functionality, and the prosecution held that none of the equipment could be repaired economically in the UK. Whether it could or would be economically repaired after export (e.g., in Nigeria) was, legally speaking, deemed to be irrelevant. It was the discard test that set in motion the coordination of jurisdiction over the waste/nonwaste distinction in Benson's case.

It is here, on this distinction between waste and nonwaste, that the importance of the work of jurisdiction in the Benson case comes to the fore. The judges and juries had to engage in the practical work of deciding the ontological status—the *what* of the case—of two objects, waste and nonwaste, that go under the same rubric of discarded electronics. But they had to do so at a conjuncture of laws that build in clashing ways of enacting those two objects.

The Work of Jurisdiction and the Politics of What

Benson and his codefendants were convicted of attempting to export e-waste from the UK to the four West African countries of Congo, Côte d'Ivoire, Ghana, and Nigeria. While the circumstances of the trial entailed the national jurisdiction of the UK, they were also directly entangled with other ways of coordinating jurisdiction over the waste/nonwaste distinction. The judges and juries had to sort out (Bowker and Leigh Star 1999) these entanglements so as to come to a judgment of Benson and his codefendants. The entanglements included the work of such jurisdiction for the EU and the OECD, as well as for signatories of the Basel Convention and of the Bamako Convention. The UK and EU sort and order jurisdiction

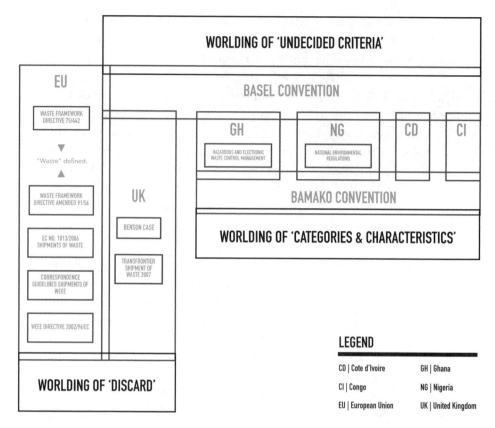

Figure 3.1
A legislative map of Joseph Benson's case.

over waste by means of the discard test. However, each of the other ways of organizing jurisdiction over waste to which Benson's case is directly attached does so differently. Indeed, there are at least three ways of worlding waste evident in the conjuncture of Benson's trial (see figure 3.1).

The copresence of these different possible ways of worlding e-waste in Benson's trial suggests three important points. First, there are multiple ways of coordinating jurisdiction over the waste/nonwaste distinction. In turn, that multiplicity tells us that no one way of worlding waste is necessary or inevitable. Second, where different worldings of waste clash, practical work will have to be done to stave off the existential challenge of that clash to the coordination of jurisdiction over the waste/nonwaste distinction. Third, managing such coordination is a provisional and contingent achievement rather than a fixed and universal aspect of being. If the latter is the case, then a key lesson is that the waste/nonwaste distinction could be organized differently, with quite different effects and consequences.

As already noted, the UK Transfrontier Act under which Benson was prosecuted invokes EC 1013/2006 for its definition of waste. It also invokes an OECD regulation for controlling transboundary shipments of waste. Regulation 23 of the UK Transfrontier Act states that a person commits an offense of the act if "he transports waste ... that is destined for recovery in a country to which the OECD Decision does not apply" (United Kingdom 2007, Article 23). All countries to which the containers from Benson's company were headed are non-OECD countries, so on this point Benson's shipments were in violation of Regulation 23. The OECD defines waste in terms of disposal, taking its wording from the original EU Waste Framework Directive of 1975 (i.e., before the EU Waste Framework Directive was amended from "dispose" to "discard"; see OECD 2008, chap. 2, Article A1). It also regulates transboundary flows of waste by means of a color-coded alert system that builds in definitions of waste from the Basel Convention. "Green" listed wastes are those "in Annex IX of the Basel Convention" (OECD 2008, chap. 2, Article B2a). "Amber" listed wastes are those in "Annex II and Annex VIII of the Basel Convention" (OECD 2008, chap. 2, Article B2b). Like the OECD regulations, EC 1013/2006 also builds into its distinction between waste and nonwaste direct references to Annex VIII and Annex IX of the Basel Convention (European Union 2006, 79).

These references to these annexes of the Basel Convention are important. Annex VIII lists materials designated as hazardous waste and thus subject to control under the Basel Convention. Annex IX also lists materials deemed waste, but because they are not deemed hazardous waste they do not come under Basel control. As discussed in the previous chapter, both Annex VIII and Annex IX enumerate materials that make up discarded electronics, making them both waste and nonwaste; meanwhile, Annex IX also permits export for reuse that "can include repair, refurbishment or upgrading, but not major reassembly," yet remains silent on what counts as "major reassembly" (Basel Secretariat 2011a, 83; see also table 2.1). In this way the question of jurisdiction over the waste/nonwaste distinction at issue in Benson's trial comprises a conjunction of laws, regulations, and treaty agreements with clashing ways of organizing the waste/nonwaste distinction. As such, judges and juries had to find a way to stave off clashes between two objects—waste and nonwaste—that go by the same name, discarded electronics.

The Basel Convention sorts and orders its jurisdiction over waste differently than the UK and EU do. Like the OECD, the Basel Convention uses the wording of the original EU Waste Framework Directive of 1975 to define waste as that which a "holder disposes of or is required to dispose of" (European Union 1975, Article 1(a)). Yet Basel also relies on a combination of categories of materials, and on the characteristics of those materials that make them hazardous (Basel Secretariat 2011a; see Annex I and Annex III therein). As I discussed in the previous chapter, the clash over the distinction between waste and nonwaste electronics in the draft technical guidelines is a clash of worldings. Judges and juries in the Benson case composed their own worlding of electronic waste through the work of jurisdiction. But that worlding clashes with the other ways of enacting electronic waste articulated in the Basel Convention. The previous chapter detailed the breakdown of negotiations over the criteria that would decide the distinction between waste and nonwaste electronics. The Basel Secretariat passed on an interim basis the technical guidelines on waste versus nonwaste electronics despite a lack of agreement about the underlying criteria used to make that distinction. Under such interim conditions, characterized by strongly polarized disagreement, passing the technical guidelines generated a "worlding of undecided criteria" (see figure 3.1).

The worlding of waste effected in Benson's case also clashes with the waste/nonwaste distinction in the Bamako Convention, a regional treaty among African nations signed by all the countries to which the containers in Benson's case were headed (Nigeria and Ghana enacted their own national laws on e-waste in 2011 and 2016, respectively; both pieces of legislation came into effect years after the events for which Benson went to trial). The worlding of categories and characteristics composed by the Bamako Convention stands in contrast to both the worlding of discards enacted in the Benson case and the worlding of undecided criteria brought into being through the technical guidelines of the Basel Convention (see figure 3.1). Bamako defines waste using language identical to that of the Basel Convention. For example, Bamako duplicates the Basel Convention's Annex I list of waste categories (e.g., clinical wastes, wastes from the production, formulation, and use of resins, latex, plasticizers, glues/adhesives) and its Annex III list of hazardous characteristics (e.g., explosive, flammable, toxic). Under Bamako, just as under Basel, wastes are "substances or objects which are disposed of, or are intended to be disposed of, or are required to be disposed of by the provisions of national law" (Bamako Secretariat 1991, 3). In this way, both Basel and Bamako derive their definition of waste from the original European Waste Framework Directive of 1975 before its amendment from "dispose" to "discard," as discussed earlier. Bamako then goes on to further specify the meaning of waste and hazardous waste under its auspices in two annexes, both of which are derived from the Basel Convention with a few modifications. Annex I of the Bamako Convention is a list of categories of wastes deemed to be hazardous. These categories are identical to those found in Annex I and Annex II of the Basel Convention except that Bamako adds an additional category for wastes containing or contaminated by radionucleotides resulting from human activity (Basel arrogates control of such substances to other extant international agreements).

Yet Bamako does not merely copy or recapitulate Basel (Kummer Peiry 1992). Unlike Basel, Bamako forgoes listing specific materials as wastes and thus avoids the internal contradictions found in Annex VIII and Annex IX of the Basel Convention, which respectively define which materials will and will not count as hazardous wastes (see table 2.1). Bamako's Annex II uses characteristics to define whether a given specific material will be hazardous. Consequently, Bamako avoids at least some of the definitional problems attendant on the distinction between waste and nonwaste that are evidenced in the failed negotiations of the technical guidelines under the Basel Convention. Bamako relies on a combination of categories (Annex I) and characteristics (Annex II) to world wastes under its auspices (see figure 3.1).

Bamako is also in some ways much more radical than Basel in scripting its jurisdiction over waste. As Zsuzsa Gille argues, "If waste production—that is, how much and what kind of wastes can be produced—is excluded from public discourse, the most that democracy can achieve is to regulate what to do with wastes already produced" (Gille 2007, 210). The treaty among African states contains language directed at promoting clean production methods, not just the control of waste already made. As Bamako states:

The Parties shall cooperate with each other in taking the appropriate measures to implement the precautionary principle to pollution prevention through the application of clean production methods, rather than the pursuit of permissible emissions approach based on assimilative capacity assumption. ...

Clean production shall not include "end-of-pipe" pollution controls such as filters and scrubbers, or chemical, physical or biological treatment. Measures which reduce the volume of waste by incineration or concentration, mask the hazard by dilution, or transfer pollutants from one environment medium to another, are also excluded [from the meaning of clean production]. (Bamako Secretariat 1991, 7)

The Basel Convention, in contrast, contains no wording at all on clean production or the precautionary principle. In these respects the work of jurisdiction in Bamako provides much more radical grounds than does Basel for imagining and enacting the necessary conditions for actually preventing, rather than merely mitigating, the production of waste from electronics.

Whether the Bamako Convention actually achieves its stated principles of precaution and clean production is, of course, a debatable point. But a much more important issue is at stake. It is this: Bamako attempts to institute jurisdictional oversight over how that which will eventually become waste is made in the first the place, that is, jurisdiction that folds manufacturing into its purview. The UK law under which Benson was charged, as well as the other legal frameworks with which his case is entangled (the EU's, the OECD's, and the Basel Convention), limit their jurisdictional purview to waste already produced. Bamako provides the jurisdictional language that would extend public decision making into the very production processes that deliver commodities that at some point later enter waste streams. The contrasts between these worldings of waste point to a crucial lesson about the organization of jurisdiction over the waste versus nonwaste distinction—it could always be organized otherwise and in ways with very different consequences for how waste comes to be. These are themes to which I return in chapters 6 and 7.

* * *

The practical work of drawing the line between waste and nonwaste at the center of Benson's legal dispute brings into being explicit and implicit goods that are sought and bads that are fought (see table 3.1). One good relates to liability in the case. The court found Benson's case to be "wholly different to a situation where a largish company, perhaps having a Board of Directors sitting somewhere," is charged with corporate malfeasance related to waste trading; thus it was justified to apply strict liability and make Benson personally liable for the offenses (Judge Dawson 2014, 6).

Another good is the coordination of EU member states' approach to waste management. This good fights the implicit bads of a patchwork regulatory environment and the ensuing commercial uncertainty such a patchwork might generate. A third good noted by the appeal judges entails certain forms of recovery, such as recycling, reuse, and repair. So long as these recovery operations occur within the EU or, subject to some conditions, within the OECD, then recovery is good. A fourth good sought is managing hazardous waste arising in an EU state. The appellants in the case were indicted with transporting waste "destined for recovery in Nigeria, a non-OECD country," and the appeal judges note that "one of the objects of the [Basel] Convention was to address the problems caused by the export of waste to developing countries" (Lord Justice Pitchford 2012, 10 para. 2). The latter goods entail another, weightier one: protection of the environment and society "as a whole," of "our children and grandchildren," of "poorer countries," and of "the world" (Judge Dawson 2014, 3–4, 10, 12).

These goods and bads are set up as such through recourse to the action of discarding. The judges and juries in Benson's case had to find a way to handle the built-in but unresolved ambiguities of the laws on the distinction between waste and nonwaste deemed relevant to Benson's prosecution. They did so by deferring to the action of discarding rather than to the categories and characteristics of Annex VIII and Annex IX of the Basel Convention, even though those annexes are explicitly part of the UK Transfrontier Act under which Benson was charged. In this respect, deference to the EU's Waste Framework Directive and Advocate General Jacobs's ruling that instituted the discard test provided a way to avoid the ontological clash inherent in the UK's jurisdiction over the waste versus nonwaste distinction.

The deference to the action of discarding demonstrates an important point. It shows that the line between waste and nonwaste could have been drawn differently. Within the case itself there was the possibility of making the distinction based on the approach of the Basel Convention, which uses categories of materials and characteristics that make them hazardous. Instead the distinction between waste and nonwaste was made using the discard test. Other ways of making the same distinction are imaginable. For example, it could have been made on the basis of what the holder of the equipment intended to do with it, such as have it repaired in the country of export. The question of intent was ruled to be an unsatisfactory

Table 3.1
Summary of goods sought and bads fought in Joseph Benson's trial for international waste trading

Goods	Bads	Source
Strict liability	Corporate malfeasance	(Lord Justice Pitchford 2012, 1 para. 3)
Coordination of EU member states' approach to waste management	Patchwork regulatory environment; commercial uncertainty between jurisdictions	(Lord Justice Pitchford 2012, 5 para. 9)
Recovery via recycling, reuse, repair, or reclamation within the EU or OECD	Recovery via recycling, reuse, repair, or reclamation in non-OECD countries when equipment is not tested for functionality	(Lord Justice Pitchford 2012, Ezeemo, 7 para. 13, 8 paras. 16–17, 10 para. 21, 22–23 paras. 57–60) (Dawson 2014, 3)
Managing hazardous waste generated in an EU state within the EU region	Export of hazardous waste to non-OECD or "developing countries" for management	(Lord Justice Pitchford 2012, 1 para. 2, 5 para. 10)
Protection of environment and society as "a whole," of "our children and grandchildren," of "poorer countries," and of the "the world"	Harm to environment and society as "a whole," to "our children and grandchildren," to "poorer countries," and to the "the world"	(Dawson 2014, 3–4, 10, 12)

way of determining what the equipment in Benson's possession was (waste or nonwaste), but that does not mean it is impossible to use that criterion. Another way to make the distinction could have been to use the location at which equipment was deposited (e.g., civic amenity sites versus retailer locations). None of these other possible ways of determining jurisdiction over the distinction between waste and nonwaste can be said to be superior to how the determination was made in Benson's case. The point is merely that the determination *could* be made differently. In this sense, neither waste nor nonwaste merely *is*. Rather, both are enacted. They are practical achievements riddled with provisionality and contingency. As such they could be enacted differently and, perhaps, in ways in which people in Joseph Benson's type of work would be found on the side of the line where goods rather than bads are seen to be being done.

Conclusion

Two key points were at issue in Benson's original trial: whether the jury was certain the items found in the containers en route to West Africa were in fact waste, and whether the jury was certain that nothing sufficient had been done by the collectors (Benson and colleagues) to change the status of the items from waste to nonwaste. Judge Black instructed the jury members that if they were "unsure" on either or both points, then their decision "must be not guilty" (see Lord Justice Pitchford 2012, para. 55). The guilty verdict delivered in the original trial (and upheld on appeal) tells us that the jury was certain on both points. But as I have argued throughout this chapter, this certainty is provisional and contingent rather than relying on an absolute or essential character of the equipment in question.

At the very heart of Benson's case are three different regulatory ways of worlding electronic waste, all of which clash. There is also Benson's own attempt at worlding electronic waste through export for repair. To the extent that the work of jurisdiction prohibited Benson's attempt, this other way of worlding ceased to subsist. Clash is avoided only by privileging one such worlding over the others. Privileging the discard test to distinguish waste from nonwaste is a practical achievement, but not one that escapes fundamental indeterminacies built into the various attachment sites that compose the case. The law under which Benson and his codefendants were charged, the UK Transfrontier Act, connects it to regulations of the EU and the OECD, the Basel Convention, and the Bamako Convention. In turn, the OECD, the Basel Convention, and the Bamako Convention offer definitions of waste partly derived from the EU's Waste Framework Directive of 1975. The legal certainty of the judgment achieved in Benson's case is one characterized by circularity and the constant bracketing out of other equally legitimate ways of worlding electronic waste at the very heart of the case. If, for example, the criteria of Annex IX of the Basel Convention, which permit export for repair, had prevailed, it is possible that Benson and his codefendants would have won rather than lost their appeal. Or indeed, they may never have been found guilty when first charged.

The judgment in Benson's case rendered jurisdiction over waste/nonwaste into a coherent whole comprised of interlegal links between

at least 186 countries, depending on how one lumps or splits membership among the EU, the OECD, and signatories to the Basel Convention
and the Bamako Convention. Yet this "global" legal geography is not
inherently big. It is generated through specific legislation and regulation
that gathers together several sites, but nothing so much as a single coherent global e-waste problem that is always and everywhere the same. The
"global" problem of e-waste is found in the noncoherent entanglements
that constitute the particulars of the attachment site that is Benson's case,
and in how those particulars travel and connect one site with another.
Noncoherence is not the same thing as incoherence. The former is, in
principle, at least partially knowable, whereas the latter fundamentally
escapes knowing. Each way of worlding electronic waste relevant to Benson's case is comprehensible, but each is also mutually exclusive of the
others. The case is noncoherent through and through; the connections
that compose it are patchily distributed in sites elsewhere beyond the
happenings in the courtrooms at Basildon, where Benson was originally
tried, and Snaresbrook, where his appeal case took place. But those connections are particular and specific. They do not extend over an abstract
global space. Benson's case links together Rue du Fort Niedergrünewald in
the city of Luxembourg, where the Court of Justice of the European Union
sits, and 2 Rue André Pascal, Paris, where the OECD is headquartered. It
connects the headquarters of the Secretariat of the Basel Convention at
11–13 Chemin des Anémones in Châtelaine, Switzerland, with the Basildon and Snaresbrook courts, UK, and with UNEP head offices on United
Nations Avenue in Nairobi, Kenya, which acts as secretariat for the Bamako
Convention. As an attachment site, Benson's case indexes a specific collection of places where change directed to reducing waste by instituting
clean production might be advocated for. What worldings of waste might
emerge if the EU, OECD, or Basel legislation and regulations against waste
adopted a stance like that of the Bamako Convention, which favors clean
production?

The legislation deemed relevant by the judges at Benson's trial takes as
the only thing within its purview waste ready-made. The legislation brackets out having any jurisdiction over how that which will be discarded is
manufactured in the first place. In this sense, the goods sought and bads
fought are set up as such by limiting public decision making to questions
about the right way to handle waste only after it already exists. As such,

the legislation brackets out of its purview questions about alternatives for reducing or even eliminating waste before it is ever generated (O'Brien 1993). Yet it is exactly these sorts of questions that have to be publicly broached—and answered—if what Gille (2007, 214) calls a "safer material world" is to be instituted. Criminalizing small-time traders like Benson who recognize the pent-up value in electronics discarded in the UK that can be reused and repaired abroad is no solution to the vastly more significant tonnage, toxicity, heterogeneity, and harms of waste arising in resource extraction for, and the manufacturing of, electronics.

4 Charting Flows of Electronic Waste

The geography is elsewhere.
—Donna Haraway (1991a, 4)

Introduction

This chapter is about where e-waste goes when it is traded internationally. But in providing answers to that seemingly simple question, it raises again the intertwined questions explored throughout the book. What counts as waste? To whom? Where? Under what conditions?

My central claim in this chapter is that what better explains the historical and emerging patterns of trade in and traffic of discarded electronics is not trade for dumping but trade for reuse, and the dynamic ecologies of repair, refurbishment, and materials recovery that drive that trade. Elizabeth Grossman's (2006, 177) description in her book, *High Tech Trash*, of the "murky channels" of "e-waste travels" was an early inspiration for my own attempts to map global flows of e-waste. In attempting to do so, I offer more up-to-date quantitative data, as well as a qualitative fieldwork-based analysis of worlds of repair and refurbishment in cases from the global south, an area that receives no detailed attention in Grossman's book except as a site of "primitive recycling." My argument is not that disposal of end-of-life electronics or the toxicological consequences thereof never occur. Instead, it is that when disposal does occur, it is typically after multiple rounds of reuse, repair, refurbishment, and materials recovery in which domestic consumption, rather than foreign dumping, plays a crucial role. As a consequence, the character, distribution, and scale of harms and benefits arising from global flows of e-waste cannot be sufficiently understood if the dominant story line about waste dumping remains the unquestioned starting point for specifying discarded electronics as a waste management problem. Said differently, my proposition is that when waste from electronics is framed predominantly as a postconsumer waste management problem and, as such, a problem captured by the dumping story line, then calls for action and policy proscriptions misspecify the issue and proffer solutions to it that are guaranteed to fail, even on their own terms.

Methods for Charting the Global Trade in and Traffic of Electronic Waste

E-waste flows are notoriously difficult to quantify and to map. There are several reasons for this. First, no universal definition of e-waste exists. This lack of universal definition is true among customs authorities globally, but it is also true in places where subnational territories have jurisdiction over electronic waste management. In the United States, for example, different states count different kinds of equipment as covered by state e-waste management legislation. As a consequence, what is e-waste in one state with such legislation may not be in another (see Lepawsky 2012, esp. table 1). Second, there do not exist trade data categories that would fully capture all meanings of the term "e-waste." This is a fundamental problem in definition that any research on the trade in and traffic of e-waste must confront, and is not a unique problem with the data I analyze here (see also Kahhat and Williams 2012; Lepawsky 2015a, 2015b). To better understand this issue of data categorization for e-waste, we may consider the ubiquitous but uneven distribution of electronics in all sorts of products. Automobiles and mechanized farm vehicles are good examples. They have become so packed full of electronics that manufacturers have argued that their engines should be covered by the U.S. Digital Millennium Copyright Act (DMCA), which would make it illegal for owners to open, tinker with, or repair their own vehicles because of the software running as part of normal engine operations (this move was partially defeated; see Wiens 2015). Vehicles are, of course, only one example of a product that might not intuitively seem "electronic" in the sense typically conveyed by the term "e-waste." If you look around your environs right now, chances are high that you are in close proximity to more than one such piece of equipment—the compact fluorescent light bulb behind me has an internal circuit board in the plastic case holding the wiring (it also contains mercury

in the bulb itself, making this "energy-saving device" also legally categorized as hazardous waste, but not e-waste where I live). These and numerous other examples illustrate the pervasiveness of electronics in ways that
overflow the meaning typically ascribed to e-waste. Such overflow makes
for a challenging data categorization problem. Third, no international
trade data sets directly capture trade for repair, refurbishment, or reuse of
electronics; and fourth, by definition, official trade statistics miss unreported trade (whether through error or deliberate practice to hide illicit
trade). But again, this is a problem that plagues global trade data sets in
general, not just those data I analyze in the next sections of this chapter.

The data set for mapping international flows of e-waste I discuss here
is built with publicly available data from the United Nations Commodity
Trade Statistics Database (COMTRADE). These data are for what is technically called Harmonized System (HS) 2002 code 854810 (waste and scrap
of primary batteries and electrical accumulators) for the years 1996–2012.
The timeliness of data reported by individual countries to the COMTRADE
database varies. Available data often lag by one or more years the current
calendar year. The period 1996–2012 is currently the date range for which
the most comprehensive data are available for HS 2002 code 854810. These
data cover some two hundred territories and more than nine thousand
import transactions. Import transactions are those trades reported by a
given territory to have been sourced from other territories. In other words,
the import data are what territory X reports coming to it from territories
Y and Z. In the case of one territory, China, the data set I use relies on
exports reported by all other territories as destined for China. This is necessary because China's imports reported to the COMTRADE database suspi-
ciously drop to zero after 2002, the year that country promulgated
legislation banning the import of e-waste. As a consequence, I assume that
the sudden change for imports reported by China likely reflects deliberate
misreporting. Relying on exports to China reported by all other territories
mitigates this possibility of deliberate misreporting and does not signifi-
cantly modify the overall patterns of observed trade.

Reported import figures are used based on the assumption that a given
territory is more likely to accurately report on what comes into its borders
than on what leaves them. It seems especially unlikely, for example, that
an importing territory would falsely report import data if such false reporting pointed to trades that violated provisions of the Basel Convention. Yet

such trade transactions can be identified. In other words, importing countries do report obtaining e-waste from sources that, under certain circumstances, are illegal under the Basel Convention. This is one reason the
COMTRADE data set, for all its limitations, is more useful for further
studies of illicit trade than might otherwise be assumed.

Once the COMTRADE data were obtained, each territory was appended
with additional information useful for subsequent analyses. The additional
information for each territory includes its status under Annex VII of the
Basel Convention (the Ban Amendment, though not yet in force, seeks to
prohibit trade from Annex VII territories to non–Annex VII territories),
whether it is a member of the Organisation for Economic Co-operation
and Development (OECD), a regional categorization (e.g., Africa, the Americas, Asia), and geographic coordinates (latitude and longitude) to permit
cartographic representations of the data set. The appended data set was
then imported into Gephi, a network analysis software with cartographic
capabilities (Gephi Consortium 2013). The resulting visualization is a car-
togram that depicts flows of e-waste between territories (see figure 4.1
online). The size of a given territory is scaled to reflect the number of other
territories reporting imports from it. The widths of the flows are scaled to
reflect the relative weight in kilograms of a given flow compared to all
other flows in the trade network.

Global Synopses, 1996–2012

The COMTRADE data indicate that between 1996 and 2012, global trade
in e-waste evolved in several distinctive patterns. The most obvious pattern
is an overall rise in total flows. Between 1996 and 2012, total global e-waste
flows more than doubled, from over 488 million kilograms to more than
1 billion kilograms, a compounded annual growth rate of about 5 percent.
Much more important than total growth in flows is how those flows are
organized. As already discussed in chapters 2 and 3, the Basel Convention
divides the globe into two blocs of countries. One bloc, known as Annex
VII, consists of members of the European Community, the OECD, and
Liechtenstein. The other bloc, known as non–Annex VII, consists of all
other signatories to the Basel Convention. Except under specific circumstances (discussed in chapters 2 and 3), and if the Ban Amendment were
in force, the Basel Convention would prohibit flows of hazardous waste,

including e-waste, from Annex VII countries to non–Annex VII countries. Given the dominant story line that e-waste flows from the developed world or global north of Annex VII territories to the developing world or global south of non–Annex VII territories, one would expect such patterns to be evident in the COMTRADE data. However, the patterns that emerge from these data differ substantially from that expected picture.

Instead of growing volumes of e-waste flowing from the rich, developed countries of the Annex VII group to the poor developing countries of the non–Annex VII group, quite different patterns of regionally organized trade are apparent (see figure 4.2). Between 1996 and 2012, e-waste moving from Annex VII countries to non–Annex VII countries shrank substantially. In 1996, about 47 percent of e-waste originating in Annex VII (or developed) countries flowed to non–Annex VII (or developing) countries. But this flow shrank quickly and remained below 7 percent of total Annex VII trade after 1998. By 2011, trade from Annex VII countries to non–Annex VII countries accounted for less than 1 percent of flows of e-waste originating in Annex VII countries. The COMTRADE data show that the vast majority of e-waste flows originating in Annex VII territories flow to other Annex VII territories.

Patterns of non–Annex VII e-waste flows differ markedly, however, from those of Annex VII. Over the 1996–2012 period, e-waste originating in non–Annex VII countries predominantly flowed out of that bloc to Annex VII countries. The overall trend shows the non–Annex VII bloc to be increasingly externally oriented; that is, growing volumes of e-waste are flowing out of the region to Annex VII countries (see figure 4.2). In short, two broad patterns are evident in the COMTRADE data over time: the Annex VII bloc is predominantly internally oriented (i.e., member countries trade among themselves), while the non–Annex VII bloc is predominantly externally oriented (i.e., trading to Annex VII territories).

In 1996 the top ten importers of e-waste included six OECD countries (France, Mexico, Belgium-Luxembourg, Sweden, the UK, and Germany). Indonesia, a non-OECD country, was by far the single largest importer of e-waste that year. Yet even so, the pattern of trade is not a monolithic one of developed countries sending e-waste to poorer, developing ones. Just over 62 percent of Indonesia's imports came from Annex VII territories in 1996 (see figure 4.3). Australia, an OECD member and thus an Annex VII territory, was the largest single source of Indonesia's e-waste imports at that

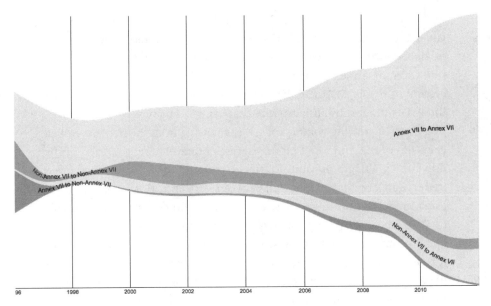

Figure 4.2
A visualization of the flows of e-waste within and between Annex VII and non–Annex VII territories over time, 1996–2012. The *y*-axis is proportionate to total flows. The *x*-axis shows time in years.
Source: Author's calculations of COMTRADE data.

time, sending over 52 million kilograms of e-waste to Indonesia in 1996. Yet second-place Hong Kong, which sent more than 35 million kilograms to Indonesia, is a non–Annex VII territory. Hong Kong's 1996 GDP per capita at U.S. $24,435 (even higher than Australia's at U.S. $19,183) was more than twenty-eight times Indonesia's in the same year. Indeed, of the non–Annex VII territories from which Indonesia's e-waste imports arrived in 1996, eleven of them had a higher GDP per capita than Indonesia and seven had a lower GDP per capita than Indonesia. Thus, available trade data and measures of wealth paint a mixed picture of e-waste imports to Indonesia in 1996. The majority of its e-waste imports did come from Annex VII territories, but this was far from being exclusively the case. With over 37 percent of imports reported by Indonesian authorities as having arrived in the country from other non–Annex VII territories, trade between territories of the non–Annex VII bloc was significant even in 1996. Moreover, the group of non–Annex VII territories from which Indonesia reported importing e-waste in 1996 was fairly evenly split in terms of wealth relative to Indonesia.

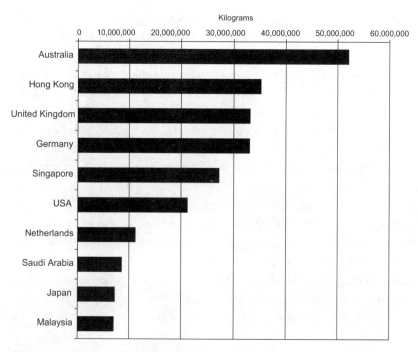

Figure 4.3

Top ten sources of e-waste imports into Indonesia, 1996.
Source: Author's calculations of COMTRADE data.

It is worth emphasizing here that if the e-waste problem is about the transboundary shipment of hazardous waste from developed to developing countries, then even at this early point in the story, such a framing of e-waste accounted for less than 50 percent of the problem (recall that in 1996, about 47 percent of e-waste originating in Annex VII countries flowed to non–Annex VII countries). By 2012 the situation was dramatically different: all the top ten importers of e-waste were Annex VII territories. Top-ranked Mexico reported imports of more than 388 million kilograms, with all but a tiny fraction coming from the United States. However, both the United States and Mexico are OECD countries, so both fall into the Annex VII bloc of the Basel Convention. Being members of the Annex VII bloc permits such trade between these two countries under the Basel Convention (though that such trade *is* permissible is not the same thing as judging that it *ought* to be so). Second-place Republic of Korea, also an OECD country, reported imports of more than 211 million kilograms. Almost 55 percent of that came from just two other OECD countries, Japan and the United States, at about 37 percent and 18 percent, respectively.

In terms of total imports of e-waste, the top-ranked non–Annex VII territory was India, in twelfth place. And yet India's situation highlights the radically different patterns of trade that emerged between 1996 and 2012. Fewer than 3 percent of India's total e-waste imports came from Annex VII territories in 2012, while more than 97 percent of India's e-waste imports—representing more than 20 million kilograms—came from other non–Annex VII territories. The largest regional source of India's e-waste imports in 2012 was Africa, with Nigeria and Angola being the top two source countries, together accounting for more than 34 percent of India's reported e-waste imports. When these broad patterns just described are disaggregated, the patterns that emerge suggest a need to completely rethink the usual framing of the e-waste problem. The most important aspect of these patterns is that they indicate a far more varied geography of trade in e-waste than the prevailing representation of the e-waste problem permits.

* * *

I am standing just inside the threshold of a small shack made mostly of wood in the middle of Agbogbloshie, a notorious e-waste site in a section of Ghana's capital city, Accra. Before arriving I had expected to be affronted by plumes of chemical smells emitted by dead electronics immolated for their copper and other metals. The plumes are here, and patches of ground are blackened from the burning, but the wind is trending away from where I stand, so the odors are faint. It is May 4, 2011, a Wednesday, warm but not yet hot. Just another day in Agbogbloshie.

Inside the shack, two young men are at work. They are loading an operating system onto one of the computers they have built for sale. The two men mostly ignore me, perhaps absorbed in their day's work or perhaps less than impressed by yet another foreigner visiting the site. Their body language suggests they have no interest in being photographed—unlike one of the young men outside, seated on an upturned CRT monitor and fixing a bike, who waved me over and posed for my camera.

The interior of the wooden shack has the look of a digital tinkerer's workshop. A clutch of CPU fan assemblies is arranged along a shelf on one wall. Their guts of yellow, red, and black wires hang in loose bundles waiting to be assembled into desktop towers. On a crowded wooden desk in front of the two men sit two CRT monitors. A black one displays a video signal test pattern. A white one charts the progress of the operating system upload. A shelf below is clogged with sixteen CD-ROM drives, grubby with past users' fingerprints. They too wait their turn to be bundled into the computers the two young men will build today. These two technicians are assembling computers entirely from the disparate bits and pieces brought here by the myriad door-to-door collectors who collect the city's digital discards and bring them to Agbogbloshie (Akese 2014). I do not have a clue how to put all these things together and make a working computer. Do you?

* * *

Africa is a net exporter of e-waste. COMTRADE data for 2012 show African territories imported almost 108,000 kilograms from other regions. But African territories were reported to be the source of almost 37 million kilograms of e-waste imported into other regions around the world. About 75 percent of those flows from Africa are imported into Asia. India, followed by the Republic of Korea, are the largest recipients of these imports from Africa. India reports receiving more than 15 million kilograms of e-waste from African sources. Nigeria, Angola, and Congo are the top three sources of India's e-waste imports from Africa; together they are the source of more than 54 percent of imports into India from the region. The Republic of Korea reports imports of more than 12.5 million kilograms of e-waste from Africa. Almost half of those imports come from Ghana. Indeed, the single largest African source of e-waste imported by Asia is Ghana, reported to send more than 6 million kilograms to countries in Asia. South Africa, at over 5.4 million kilograms, and Nigeria, at just under 5 million kilograms, are the second and third largest African sources of e-waste imports into Asia. After Asia, countries of the Middle East and Europe report the largest imports of e-waste from Africa. Israel alone accounts for nearly 18 percent of total imports reported from African sources (more than 6 million kilograms arrive from South Africa and more than 102,000 kilograms from Morocco). Europe accounts for the remaining approximately 6.5 percent of total imports reported from Africa. The two most important African sources of European imports are Nigeria, at over 1.2 million kilograms, and South Africa, at over 965,000 kilograms. The top European importer of e-waste from Africa is Belgium, which brings in almost 2.1 million kilograms, with about 60 percent of those imports reported to come from Nigeria.

The figures reported above may surprise readers familiar with the dominant story line about e-waste. In that narrative, Africa is always the destination of e-waste flows. It is, to paraphrase a recent headline, the graveyard of the world's e-waste (Akbar 2015). Owing to the inherent limitations of the data, as discussed above, one might question the accuracy of findings derived from COMTRADE data. And yet a growing body of research drawing on different sources of data and using different methods points in the same direction as the COMTRADE data. One source of this evidence is the United Nations' E-Waste Africa Program, which ran between 2008 and 2012 (Secretariat of the Basel Convention 2011). The program's summary report, *Where Are WEEE in Africa?*, assumed that "in 2010 between 50–85% of e-waste [was] domestically generated out of the consumption of new or used EEE [electrical and electronic equipment]" (Schluep et al. 2011, 10). In other words, e-waste arising domestically in the African countries studied was the source of at least half of the total e-waste arising in those countries. Imports of e-waste, on the other hand, accounted for the other 15–50 percent.

While ENGOs and reportage focus on cheap labor and low environmental standards to explain the geography of e-waste flows in terms of a supply push, Bloomberg journalist Adam Minter (2013a) shows what is missing from such assessments: the demand pull generated by the economics of shipping, as well as the presence of sufficient infrastructure and know-how for putting back into use what others have discarded as waste. When containers of new electronics (and other commodities) arrive in wealthy markets and are emptied, if little or nothing fills them back up for the return journey, prices for space on the backhaul go down. But low shipping prices for backhaul journeys from, for example, the Port of Los Angeles to the Port of Yantain, China, are not enough on their own to shape the geography of trade in discards. There must also be the industrial infrastructure and requisite skill base to turn what others treat as waste into valued commodities. This is why Yantain's proximity to Shenzhen makes that port a crucial node in the network of both scrap exports from abroad and the so-called "Workshop of the World" that exports new manufactures to consumers (*Economist* 2002). Indeed, as Minter (2013a, 94) points out, despite Sudan having desperately low labor costs and environmental standards, it is nevertheless an exporter of scrap to India and China (see also Minter 2016).

Countries in West Africa, particularly Ghana and Nigeria, have been made notorious in ENGO and media reports as "hot spots" of e-waste imported from countries of the global north (e.g., Blacksmith Institute 2013; Höges 2009; Klein 2009; Puckett et al. 2005). Yet the E-Waste Africa reports outline quite a different picture of the situation overall and in the specific countries studied. The Nigerian reports (Manhart et al. 2011; Ogungbuyi et al. 2012) used a mixed methods approach, combining both quantitative and qualitative data. They included data for new and used electronics from shipment manifests, export data from European sources, data from the Nigerian National Bureau of Statistics and the Nigerian Customs Service, and interviews with the National Port Authority, customs authorities, and importers, in addition to fieldwork results. Manhart and co-workers (2011, xii) write that "70% of all the imported used equipment is functional and is sold to consumers after testing. 70% of the non-functional share can be repaired within the major markets and is also sold to consumers. 9% of the total imports of used equipment is non-repairable and is directly passed on to collectors and recyclers." In other words, 91 percent of total imports of used equipment are functional or repairable.

A second study of Nigeria from the E-Waste Africa project found that "approximately 30% of second-hand imports were estimated to be non-functioning (therefore need to be declared as e-waste): half of this amount was repaired locally and sold to consumers and the other half was un-repairable. In the case of Nigeria, this was at least 100,000 tonnes of e-waste entering the country illegally in 2010" (Ogungbuyi et al. 2012, 3). The claim that the portion of the imported used electronics that is nonfunctioning must be declared e-waste is actually part of an ongoing debate within the Basel process itself (see chapter 2). For the moment it is the straight arithmetic of this report on Nigeria I want to draw attention to. The report's own findings mean that 70 percent of secondhand imports are functioning. It also means that of the 30 percent that are nonfunctioning, half were repaired and sold locally, leaving 15 percent of total second-hand imports, or 90,000–126,000 metric tons, as nonfunctioning "waste" imported from abroad. Meanwhile, the report estimated that "around 440,000 tonnes" of the total domestically installed base of electronic equipment in Nigeria "ends up as e-waste" annually (Ogungbuyi et al. 2012, 3). By the report's own estimates, then, e-waste arising from Nigerian household, institutional, and corporate consumers—that is, from domestic sources—is 3.5 to 4.8 times greater than e-waste imports from abroad. Some might argue that even if imports of e-waste are less than what arises domestically, that is no excuse for the dumping of e-waste in Nigeria by richer countries abroad. There is merit to that claim. But if the dominant way of framing e-waste as a problem of environmentally sound management relies on narrating that problem as a story about dumping from abroad, then at least in the case of Nigeria we are stuck deriving solutions for only about 15 percent of the problem. Shutting off imports of e-waste to Nigeria—even assuming this were possible or a desirable way of handling those imports—would do nothing to mitigate the 440,000 metric tons of e-waste estimated to be generated domestically on an annual basis in the country.

Another report from the E-Waste Africa program covers Liberia (Strother, Williams, and Schluep 2012). It uses a mixed methods approach similar to that used by Ogungbuyi and co-workers (2012) in their study of Nigeria. Strother and co-workers (2012) combined what few statistical data were available with interviews with customs authorities and with results of fieldwork at disposal sites and sites of repair. The authors were careful to point out the limitations of their study (among them, having to rely on assumptions because of the lack of formal data collection about the sectors involved). But again, the picture that emerges contrasts in important ways with the dominant e-waste story line. Strother and co-workers (2012, 3) estimate that a total of 3,500 metric tons of electrical and electronic equipment is imported by Liberia annually. Of that total, 10 percent, or 350 metric tons, consists of used commodities, and 70 percent of the used equipment, or 245 metric tons, arrives functioning. Meanwhile, 15 percent of the used equipment needs repair and 15 percent, or 53 tons, is estimated to enter Liberia in an unrepairable condition and is, therefore, e-waste. The report goes on to note that "approx. 90% of imports are new products" but that these have "a low life expectancy just like second-hand products," largely because they are suspected to be faked brands imported from Asia and the Middle East (Strother, Williams, and Schluep 2012, 3). The flows into Liberia include used electronics from some Annex VII states (e.g., Belgium, Germany, the UK, Japan, Korea, and the United States), but also a range of non–Annex VII states (e.g., China, Hong Kong, India, Dubai, and Egypt, as well as from the African states of Guinea and Nigeria) (Strother, Williams, and Schluep 2012; see esp. table 4, p. 14). The data in the report offer no way to apportion the relative importance of flows from this variable collection of sources, but the study offers a very different picture of flows of commodities and waste than the usual depiction of the e-waste problem.

A study of Benin (EMPA 2011; see figure 18, p. 51), again using mixed methods, estimates that 6,900 metric tons of electrical and electronic equipment are imported into the country annually. Of that total, "Second-hand equipment accounts for more than a third of total imports and are of very variable quality" (EMPA 2011, 5). At the same time, the report finds that 500 metric tons—or about 7 percent—of imports are waste (EMPA 2011, 51). Meanwhile approximately 4,100 metric tons of e-waste arise domestically in the country from households and businesses annually (EMPA 2011; see figure 18, p. 51). The study notes that the e-waste arising from households occurs after periods of use of the equipment ranging from 1.5 to eight years, depending on the category of equipment (EMPA 2011; see table 10, p. 43). At the low end of this timescale is lighting equipment and at the high end is refrigeration equipment. Cell phones are estimated to have useful lives of two years and televisions and computers five years, both of which match commonly assumed useful lives of electronics in North American markets (e.g., Kang and Schoenung 2005). If the estimate for the proportion of total imports that are "secondhand" holds at the household level, then at least one-third of the e-waste arising from house-holds does so after a round of reuse that roughly doubles the overall working life of the electronics (i.e., after "firsthand" use comes a round of "secondhand" use that appears to be of roughly the same duration as firsthand use).

The situation found for Ghana—ostensibly the site of the largest e-waste dump in the world—also differs markedly from representations of it in ENGO reports and the vast majority of media reports. In their report for the E-Waste Africa project, Amoya-Osei and co-workers (2011) found that Ghana's imports of electrical and electronic equipment totaled 215,000 tons. Of that total, about 30 percent was found to be new product and 70 percent, or approximately 150,500 tons, was "second hand." Of the sec-ondhand equipment, 15 percent (about 22,575 tons) could not be repaired and was found to be "unsellable." Put differently, 85 percent of the sec-ondhand equipment brought into Ghana is marketable, that is, not waste. Amoya-Osei and co-workers (2011, x–xi) also found that "of the 280,000 tons of obsolete devices generated in 2009 … about 171,000 tons of WEEE [waste electrical and electronic equipment] from consumers, repair shops and communal collection reached the informal recycling sector." In other words, more than 50 percent of e-waste in Ghana arises domestically and not from imports of e-waste from abroad (see Lepawsky 2015a and Lepawsky, Goldstein, and Schulz 2015 for further discussion).

Other considerable efforts by European authorities have been made to quantify flows of discarded electronics, particularly illegal shipments, that leave the EU. Between 2007 and 2013 a consortium comprised of the International Criminal Police Organization (INTERPOL), the United Nations Interregional Crime and Justice Research Institute (UNICRI), the United Nations University (UNU), and research consultants ran the Countering WEEE Illegal Trade (CWIT) project (CWIT Project 2016). Final results from this research further illuminate a disjuncture between the dominant e-waste story line and a growing body of evidence about the international trade and traffic of discarded electronics. The CWIT project found that 65 percent (or 6.15 million tons) of electronic equipment discarded in Europe is dealt with outside official collection systems. Of that 6.15 million tons, 1.5 million tons are exported out of Europe. These figures indicate, then, that about 24 percent of the total mass of material handled outside official systems leaves Europe (Huisman et al. 2015, 6, 16–17). Of the 1.5 million tons that are exported, 200,000 tons are found to be properly documented legal exports of used electrical and electronic equipment (UEEE) (Huisman et al. 2015, 16). Of the remaining 1.3 million tons, 70 percent (about 900,000 tons) are estimated to be functioning secondhand items and 30 percent (or about 400,000 tons) are e-waste but also include repairable items (Huisman et al. 2015, 16). The remaining 1.3 million tons are, the CWIT report notes, in a "grey area subject to different legal interpretations and susceptible to export ban violations" (Huisman et al. 2015, 16).

For the moment, it is just the arithmetic implications of the CWIT report that I want to draw attention to. As the report notes, the "mismanagement of discarded electronics within Europe involves ten times the volume of e-waste shipped to foreign shores in undocumented exports" (Huisman et al. 2015, 16). Moreover, CWIT's analysis of the economic drivers of exports finds that "the main driver behind exports is the reuse value combined with the avoided costs of sorting, testing, and packaging" of discarded electronics and other electrical equipment (Huisman et al.

2015, 18). Indeed, "the magnitude of reuse value is multiple times the material value" of exported discarded electronics (Huisman et al. 2015, 18). In other words, trade for reuse rather than export for waste dumping is found to be the main economic driver of such exports. Here again, we are presented with a picture of e-waste flows that jars badly with the dominant story line, which casts the problem as one of waste dumping from rich countries to poor ones.

While the CWIT report quantifies exports from Europe, it does not specify in detail where those exports go. A subsequent study from the UN's Solve the E-waste Problem (StEP) initiative attempts to do just that (Baldé, Wang, and Kuehr 2016). The study analyzed quantitative data on mass and value derived from an EU trade database and employed a method developed in separate studies (see Duan et al. 2013, 2014). The StEP report found that less than 1 percent of total e-waste generated is exported from the EU (Baldé, Wang, and Kuehr 2016, 11). Meanwhile, the share of products imported from Europe compared to "waste domestically generated by the same appliances" was found to be a maximum of 16 percent in African and 6 percent in Asian destinations of those imports (Baldé, Wang, and Kuehr 2016, 17, 20). In other words, this report indicates that in Africa, at least 84 percent of e-waste arises from domestic sources rather than being imports of waste. In Asia the figure is 94 percent. The authors of the StEP study conclude their report on a note of caution. They state that their findings are "most likely an underestimation of the total amounts of exported as whole units from the EU-28" (Baldé, Wang, and Kuehr 2016, 26). But those findings would have to underestimate exports by at least two orders of magnitude to match figures about international e-waste flows that commonly circulate with the dominant e-waste story line.

* * *

It is July 29, 2009, and I am in Dhaka, Bangladesh. At the end of a long day following the flow of digital things through the city I receive an email from a journal editor. The proofs of an article on global flows of e-waste I copublished with a graduate student are ready. The journal needs a response within a few days. Eager to move the paper along to publication, I head to a nearby copy shop with a graduate student I am working with here. A key point made in this paper, which one of the shop's employees has pulled up on a screen for printing, is that "materials disposed of as e-waste in one place become sources of value elsewhere when they are reused, repurposed and/or broken down as feedstocks of primary inputs" (Lepawsky and McNabb 2010, 191).

The manuscript appears in front of us on a Philips model 107T CRT. I quickly snap a photograph (see figure 4.4). I am slightly giddy seeing my scholarly work both on the screen and working itself out in practical terms right in front of me. But, while printing out the paper, as Dhaka sweltered outside, I did not think quickly enough to ask where the shop had obtained its equipment. Was it from one of the many auctions of used electronics from banks and other institutions in the city? Was it bought from a local retailer of used electronics imported from elsewhere? Had it been bought new? Because I did not ask, I cannot say for sure what the provenance was of this specific monitor displaying the manuscript. What I can say with some degree of certainty is that the Philips Corporation began manufacturing the model 107T in the year 2000. That meant the monitor I was looking at in the copy shop might have been nine years old. In retrospect—and after much learning from fieldwork—I should not have been surprised by the scene. Only three days earlier, while we spent time with computer repair technicians clustered around one of the city's hubs for digital publishing, we were given a quick tour through a publishing business. The employees there were using Apple Power Mac G4s running OS 9 (see figure 4.5). Apple had discontinued manufacturing that computer model five years earlier, in 2004. Two years before that, the death of the OS 9 operating system had famously been eulogized by Steve Jobs at Apple's developer conference in 2002 (JoshuaG 2006).The publishing house we were being led through was running its business on a backbone of digital infrastructure declared dead five to seven years earlier in California. Over the course of several months' work we would find that used electronics imported from abroad were crucial to Dhaka's economy. We would also learn that reuse and repair are fundamental to keeping those electronics going. We are only just starting to understand how important reuse, repair, and refurbishment are to global flows of discarded electronics (Ahmed, Jackson, and Rifat 2015; Jackson, Pompe, and Krieshok 2012; Lepawsky and Billah 2011).

* * *

In contrast with Africa, the COMTRADE data show Asia to be a net importer of e-waste. E-waste flows from all other regions to Asia total just over 141 million kilograms. The net flow to Asia amounts to almost 139 million kilograms. At first the situation might look like the familiar framing of

Figure 4.4
A Philips model 107T monitor at a print shop in Dhaka, Bangladesh, in 2009 displaying the editorial proof of a journal article.
Source: Author's photograph.

e-waste being dumped in poorer countries of the global south by richer countries of the global north. Yet the underlying geography of trade is far more complex and interesting. For example, Asia is a net importer of e-waste from the Americas, yet more than 45.8 million kilograms—or more than 96 percent of those flows—consist of imports to Korea and Japan, both of which are Annex VII states. Of those flows from the Americas to Annex VII states in Asia (Korea and Japan), more than 8.4 million kilograms are imported from non–Annex VII territories in the Americas region.

Trade within the Asian region itself accounts for almost 44 percent of total trade volume for Asia. The largest importer in Asia by far is the Republic of Korea. In 2012 it reported imports of more than 211 million kilograms of e-waste. Korea is an OECD country and an Annex VII signatory to the Basel Convention. More than 63 percent of Korea's total reported imports come from just three countries: Japan (more than 78.7 million kilograms), the United States (more than 37.2 million kilograms), and the United Arab Emirates (UAE; more than 18.9 million kilograms). Japan and the United States are also both OECD and Annex VII states. The UAE is

Figure 4.5
A Power Mac G4 running OS 9 at a digital publishing business in Dhaka, Bangladesh, 2009.
Source: Author's photograph.

not. The trade flow from UAE to Korea is an example of a substantial flow of e-waste moving from a non–Annex VII territory to an Annex VII territory. Indeed, just over 38 percent of Korea's reported imports come from non–Annex VII states. A recent study by Breivik and co-workers (2014, 8736) claims, "To the best of our knowledge there is no significant export of e-waste from non-OECD to OECD." What might count as 'significant' for these authors is left unstated, but the specific instance of Korea and the broader patterns of Annex VII and non–Annex VII trade discussed above suggest the assumptions made by Breivik and co-workers (2014) are unwarranted.

India and the Philippines rank respectively as the second and third largest importers of e-waste in the region. Both are substantial net importers. Data for India show net imports in excess of 20.8 million kilograms, while for the Philippines the figure is above 4.7 million kilograms. However, in both cases more than 97 percent of these countries' total e-waste imports come from other non–Annex VII countries. For India it is African sources that dominate: just over 73 percent of India's reported e-waste imports arrive from African countries. For Philippines, the dominant source by far is Singapore, which is reported as the source for more than 7.2 million kilograms (or almost 96 percent) of e-waste imports into the Philippines. In both India and the Philippines, flows from Annex VII regions—flows which the Basel Convention seeks to prohibit—account for less than 3 percent of total e-waste flows to these two countries.

China, meanwhile, is a recurrent focus of ENGO, media, and scholarly investigations of e-waste. The hazards experienced by workers in particular segments of the discarded electronics recovery economy at sites in China are widely documented (see, e.g., Tan et al. 2016; Tao et al. 2014; Y. Wang et al. 2016; P. Xu et al. 2015; X. Xu et al. 2015; Zhang et al. 2014). However, the orthodoxy about the origins of discarded electronics that flow into China is rarely questioned. It is assumed that prevailing characterizations of global e-waste flows arriving from rich, developed countries in poorer developing countries is relevant to the situation in China as well. However, as with the African cases already discussed, there is reason to question this presumption. Baldé and co-workers' (2016) analysis reports that imports into East Asia from the EU amount to less than 1 percent of total e-waste generated domestically in that region. The authors note, "The low percentage is due to the large domestic generation of e-waste market in China" (Baldé, Wang, and Kuehr 2016, 21; see also Duan et al. 2015, figure 2; Li, Yang, and Liu 2015, figure 1).

The COMTRADE data indicate that China is a net importer of e-waste, yet even if imports reported by Hong Kong are included, the country ranks fourth after the Philippines in terms of net imports. As with the e-waste trade flows for Korea and India, the underlying geography of China's is much more complex and interesting than prevailing representations of the situation. About 64 percent of e-waste flows to China come from Annex VII territories, leaving non–Annex VII countries to account for the rest. The Netherlands at over 2.4 million kilograms is the top source of e-waste flows to China. This flow amounts to almost 50 percent of all imports into China. Brazil, Japan, Thailand, and Indonesia (beating out the UK) round out the top five sources. China is poorer in GDP per capita terms than all of the countries it imports e-waste from except for two, Nicaragua and Georgia. This situation might appear to support the prevailing image of global e-waste flows being dumped by richer nations in poorer ones. But when exports from China are examined, the picture changes. Just two countries, Japan and Korea, account for over 71 percent of e-waste imports reported to arrive from China. All told, more than 81 percent of China's e-waste exports go to countries richer than it in GDP per capita terms.

* * *

It is Friday, August 7, 2015. I am in Lima, Peru, working with a colleague and a graduate student to document electronics repair in the city. We are filming an interview with Angela, who works as a television repair technician. She works in a shop she shares with her husband. It was, she tells us, by watching him and, "more than anything, practice" (interview, August 7, 2015) that enabled her to learn to fix televisions.

The shop is small, perhaps four square meters in size. At the back is a battered set of dull gray steel shelves holding a smorgasbord of parts—circuit boards, speakers, signal converters—in stacks festooned with loose wires. Angela is at work on the guts of a large CRT. All of the housing has been removed from it. There are the electron gun, the high-voltage line, and the power supply. There are the deflection coils. If Angela were after the material value of this monitor, she would be digging in here for the copper. But she is not tearing the monitor apart. She is putting it back together. She has been doing this kind of work for seventeen years.

The people Angela and her husband do repair work for sometimes come from far outside Lima. Sometimes, Angela tells us, they undertake arduous journeys: "The jungle, Mazamari, is the furthest from which people have come. They travel by boat. It takes two to three days from Mazamari. They bring televisions to be repaired, if they can't find how to fix it there" (interview, August 7, 2015). But the market is changing, says Angela. "Most people who buy a LED television, the flat ones, they break down. And people prefer a TV that can be repaired. Here [pointing to the CRT] you can change the screen, the parts. But in the other [the flat screens], you can't change anything." As the interview winds down, Angela turns her attention back to the repairs of the CRT in front of her. A manufacturer's sticker is visible on the television's power supply housing. It was assembled in March 1995 at a factory in Tijuana, Mexico.

Conclusion

The prevailing way of answering the question of where e-waste goes when it is traded internationally is to bring up the dumping story line. Yet the patterns apparent in the COMTRADE data, along with the growing body of evidence reviewed in this chapter, strongly suggest a need to rethink the dominant approach to e-waste as a problem of waste dumping (Lepawsky 2015b). Reuse, refurbishment, and repair rarely or only peripherally inform the dominant e-waste story line. Meanwhile, dumping seemingly points to a ready solution: recycling. And it is recycling that is most commonly presented as the solution to the problem so framed. Yet recycling itself results in waste arising from discarded electronics. As the connections between electronics recycled in Roseville, California, and subsequently smelted in Quebec and New Brunswick that opened the book suggest, electronics recycling entails long-distance transport to one of the very small number of smelters—about half a dozen—on the planet that can handle electronics fractions as part of their inputs (Schluep et al. 2009, 54; F. Wang et al. 2012, 2136). Recycling in this manner means that equipment is destroyed and energy is used to do that. Emissions are released all along the way (one study finds 500 kilometers of road transport to be the upper limit electronics can be transported for recycling before the emissions impacts of that transport negate the environmental benefits of recycling the electronics; see Barba-Gutiérrez, Adenso-Díaz, and Hopp 2008, 488).

One of my goals in this chapter was to carry forward the analytical tactic of defamiliarization. I hope the discussion of international flows of e-waste invites readers to be open to reconsidering their preconceived notions of what e-waste is, where it travels, who works with it, and under what conditions. Readers should not mistake my claim to be that disposal of end-of-life electronics or the toxicological consequences thereof do not occur. When disposal does occur it is typically after multiple rounds of reuse, repair, refurbishment, and materials recovery in which domestic consumption, rather than foreign dumping, plays a crucial role. Working with e-waste to recover materials and rekindle value can be risky. But these risks inhere not only at the sites typically highlighted in accounts of the e-waste problem. Even when e-waste is recycled under carefully controlled conditions in certified facilities, workers and their families can be exposed to toxic risks from these processes (for examples in the United States, see Ceballos, Gong, and Page 2015; Newman et al. 2015).

Readers should not mistake my claim to be that access to livelihoods or technology by way of the international trade in and traffic of discarded electronics is inherently more beneficial than the toxic risks those flows entrain with them. What I am claiming is that, first, if equitable solutions to the e-waste problem are to be devised, then issues such as livelihoods, toxic risks, access to technology, pollution, learning, and enskillment need to be carefully examined together in their tangly knots and not as if they subsisted in isolation from one another.

There is a topography to global flows of e-waste. But that topography does not match the geographical imaginary of it curated by the dominant story line. As the geographer Derek Gregory writes, "Geographical imaginaries are more than representations or constructions of the world: they are vitally implicated in a material, sensuous process of 'worlding'" (Gregory 2009). The character, distribution, and scale of harms and benefits arising from global flows of e-waste cannot be sufficiently understood if the dominant story line about waste dumping remains the unquestioned starting point for specifying discarded electronics as a waste management problem.

Since the emergence of e-waste as an issue of public concern in the early 2000s, the prevailing characterization of the issue by activists, journalists, and scholars continues—almost without exception—to be one of waste being dumped by the world's rich onto the world's poor. The quantitative trade data explored in this chapter certainly have their limitations. Waste in general, and e-waste specifically, are notoriously difficult to quantify (MacBride 2012). Much data on waste generation are collected at the individual or household level and measured as municipal solid waste (MSW). Yet MSW data tell us little about waste arising in industrial production. The problem of knowing waste is partly a problem of what to count and where. When data about e-waste generation are gathered, the measurements are typically weight measurements. Weight is relatively easy to measure, but it tells us nothing about other important characteristics of e-waste, such as its potential toxicity. When e-waste generation is measured at the loci of what individuals, households, or businesses put into the municipal waste stream, the vastly larger amounts of solid, liquid, and gaseous "externalities" arising from resource extraction for, and the manufacturing of, electronics are bracketed out of consideration of what constitutes waste from electronics. Thus, to measure waste in some ways and not others is a problem of knowledge or epistemology—how we come to know what we think we know—with practical consequences: to construe e-waste as a postconsumer waste management problem suggests that practical policies such as household recycling are the right solution. Yet if most of the waste associated with electronics comes into being before consumers even purchase the items (as it does in raw materials extraction and manufacturing processes), then household recycling programs will do little to mitigate waste generated from electronics. We need to rethink the e-waste problem. How we have come to know e-waste—namely, as a matter of postconsumer dumping—is part of the problem. It downplays or even erases the attention needed to all the geographies of waste from electronics that exceed the trade in and traffic of discarded equipment. It may help to probe more deeply a seemingly simple question: How do we know e-waste? It is this question that I take up in the next chapter.

5 Looking Again in a Different Way

Indeterminacy is the condition of measuring.
—Myra Hird (2012, 462)

The etymology of the word "fact" is tricky as well as enlightening: it may mean
fabricated thus false, or fabricated thus solid. It is the second path we invite you
to follow. Cogitamus not cogito.
—Bruno Latour (n.d., 1)

Introduction

On May 9, 2016, the Basel Action Network (BAN) publicly released details
of its e-Trash Transparency Project (Basel Action Network 2016c, 2016d).
The project, done in conjunction with MIT's Senseable City Lab, surreptitiously placed 205 global positioning system (GPS)–enabled trackers in
displays and printers. These items were then delivered to e-waste collection
facilities at sites in various regions of the United States. The results of the
GPS readings come with date-time stamps down to the second and location
specified to five or six decimal places (equivalent to the level of meters).
With dazzling precision, the e-Trash Transparency Project offers a tantaliz-
ing reveal of trade routes that would otherwise have remained invisible to
those outside the electronics reuse and recycling industry (Ratti et al.
2016). What do these tracks tell us? The accompanying report shows that
sixty-five of the tracked devices (or 32.5 percent) were exported from the
United States (Basel Action Network 2016c, 8). BAN claims that "by our
best estimation and understanding of waste trade law, 62 (31% of total
trackers) of those 65 devices exported were likely to be illegal" (Basel Action
Network 2016c, 8). A summary table provided in the report shows the final

destinations of the tracked devices: Hong Kong (thirty-seven devices), mainland China (eight devices), Taiwan (five devices), Pakistan (four devices), Mexico (three devices), Canada (two devices), and Thailand (two devices), while the United Arab Emirates, Kenya, Cambodia, and the Dominican Republic each received one device (Basel Action Network 2016c, 9).

The e-Trash Transparency Project is alluring. It offers a narrative emotionally charged with concern for social and environmental justice and flavored with doses of investigative journalism and espionage cinematics, all amplified by broadcast media (KCTS 9 2016). At the same time, a reader who makes the effort to go through the entire 114 pages of the e-Trash Transparency report will also find some notes of hesitation as well as important caveats about the methods used and the findings those methods can reveal. For example, the report notes that 205 trackers are a small sample size relative to the overall volume of electronic waste generated annually in the United States. On this point the report states, "It is important, therefore, to resist arriving at sweeping conclusions on the basis of the limited data set" (Basel Action Network 2016c, 8). A few pages later, after characterizing findings about one particularly well-known electronics take-back system (Dell's Reconnect program) as "staggering" the report states, "In fact, the actual total weight of Reconnect exports could be greater or smaller" (Basel Action Network 2016c, 15). Much later in the report the authors warn that their "extrapolations must be used with caution" and present readers with a number of important caveats, worth quoting in full here:

> The extrapolations we have done in this study to indicate the potential scale of the export concern for example, must be understood with respect to a vast array of variables, which could deviate from a fair representation of the norm.
>
> Pernicious error is a danger in any study. For example, some might argue that an advocacy organization like BAN will have a built-in bias to seek out high-risk export destinations to make their case. On the other hand, the marketplace, at the time the study was conducted, had historic lows in commodity prices, and along with heightened import controls in China could have skewed the data against a robust export trade to Asia compared to even 3 years ago. The regions we chose could have for some reason not ... been nationally representative. The types of devices we chose could have been unrepresentative of the entire scope of what is generally considered e-waste. BAN, choosing to mostly focus on but one charity may have skewed the outcome, as we are not sure whether Goodwill is representative of all charities that process e-waste. Certainly BAN's study did not look into the brokers

and traders that buy directly from businesses and do not accept equipment from the general public. Nor did we survey government auction programs, which are legally obliged to seek out the least expensive methods of disposal to save taxpayer expense. There is reason to believe that brokers and government disposal is more prone to export.

As we can see, there are many variables which could skew data one way or the other. For this reason, it is important to understand that the extrapolations made in this report, based as they are on conservative estimates, are provided not as facts but as illustrative of the potentially massive scale of the problem identified. (Basel Action Network 2016c, 109)

These caveats are important, and the authors should be applauded for making them explicit. In light of these caveats and the broader sense the report gives of a large-scale crisis, however, what is a reader to make of the e-Trash Transparency report's findings?

Describing the scale and significance of their results, the authors claim that 1,256,000 U.S. tons of e-waste reach U.S.-based recyclers annually. Then, based on the GPS data, they claim that of the 1,256,000 U.S. tons of e-waste reaching U.S.-based recyclers, 25–39 percent of it is exported. By straight arithmetic, the upper-range figure of 39 percent amounts to 489,840 U.S. tons moving offshore (this is an amount even higher than the authors state on p. 10 because they include only the recycling industry and bracket out charities and retailers in their discussion of scale and significance at this point in the report. See Basel Action Network 2016c, 10). Addressing the issue of the significance of these flows, the authors note that the amounts offshored "would equate to about 18,840 [shipping] containers per annum or about 52 containers per day being exported from the US" (Basel Action Network 2016c, 10). A few pages later the report addresses its findings from trackers placed in Dell's Reconnect program. Here the authors' tone suggests crisis: "The concern over these exports is far more serious than the impact of 7 wayward electronic waste products managed by Dell and Goodwill. Extrapolation of the entire amount Dell Reconnect has handled to date, we arrive at a staggering figure of 90 million pounds exported" (Basel Action Network 2016c, 15).

Some 489,840 U.S. tons, 90 million pounds. These numbers seem big. But how big are they, and compared to what? Here is another comparison to situate these measures of weight: one *single* smelter operation in Mexico that produces copper and other metals important for electronics manufacturing generates 902,792 U.S. tons (or 819,000 metric tons) of sulfuric acid

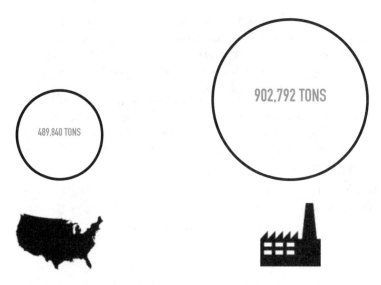

Figure 5.1
Total annual exports of e-waste from the United States compared to waste acid production at a single smelter. Circles above the icons are proportionate to the tons they represent.

annually as a waste by-product of the smelting process (Grupo México and Southern Copper 2014). If you do the arithmetic the generation of acid waste from this one smelter is about 1.8 times larger than the amount of e-waste the e-Trash Transparency report estimates is exported from the *entire* United States *annually* (see figure 5.1). It is about twenty times larger than the 90 million pounds of e-waste reportedly exported by the Dell Reconnect program. These different ways of interpreting findings such as those from the e-Trash Transparency Project or those discussed in the previous chapter raise a crucial question: How do we know e-waste?

* * *

The problem of knowing waste is partly a problem of what to count and where. Data about e-waste generation are typically measured in terms of weight. Weight is relatively easy to measure, but it tells us nothing about other important characteristics of e-waste such as its potential for toxicity (e.g., 1 kilogram of aluminum and 1 kilogram of mercury are identical in terms of weight but radically different in terms of toxicity). When e-waste generation is measured at the level of what individuals, households, or businesses put into the municipal waste stream, the vastly larger amounts of solid, liquid, and gaseous externalities arising from mining and from electronics manufacturing are bracketed out of consideration of what constitutes waste from electronics. Thus, to measure waste in some ways and not others is a problem of knowledge or epistemology—how we come to know what we think we know—with practical consequences: to construe e-waste as a postconsumer waste management problem suggests that practical policies such as household recycling are the right solution. Yet if most of the waste arising from electronics is generated before consumers even purchase the items (as it does in mining and manufacturing), then household recycling programs will do little if anything to mitigate waste generated by electronics.

Much of the rest of this chapter focuses on knowledge-making practices in a single environmental nongovernmental organization (ENGO) report and its circulation in scholarly and nonscholarly networks. That report is BAN's *Exporting Harm*. My focus on it is not without reason. The report is the source of the often repeated figure that 50–80 percent of e-waste is exported out of the United States. That figure moved out of the report and directly into legislative discussions in the United States. Also, *Exporting Harm* is one of the most widely cited documents in the academic literature on e-waste. The report illustrates the key forms of knowledge-making practices used in activist and journalistic ways of knowing e-waste, in particular documentary photography and asset tag evidence. I focus on the report for these reasons, that is, because of how people and organizations other than BAN have incorporated it into their knowledge-making practices about e-waste. My purpose, however, is less to critique the report's findings than it is to understand why the knowledge claims it makes have come to be so convincing and to have so deeply informed the dominant e-waste story line. I am also concerned with the effects such claims generate, that is, with the work they do in shaping how we know e-waste.

* * *

On February 11, 2009, members of the U.S. Congress Committee on Science and Technology heard testimony on a draft bill provisionally titled the Electronic Waste Research and Development Act of 2009 (U.S. Congress 2009, 3). The bill never became law, but before it died the committee heard testimony from five witnesses representing academia, a nonprofit trade association, manufacturers, and reuse and recycling interests. The hearing pointed to six broad areas it explicitly labeled "Issues and Concerns" (U.S. Congress 2009, 3–5). In sum (and in order of appearance), these matters of concern were that e-waste "is a growing problem in the U.S. and world-wide," that the U.S.-based recycling industry "faces a number of challenges" (including consumer participation, logistics, cost, and safety), that current design practices for electronics inhibited affordable recycling, that the chemical and material heterogeneity of electronics represented both health and environmental risks while also impeding affordable recycling, that there was a need for research "to increase and encourage re-use," and last, that there was a need to improve workforce education so that products could be designed "with a minimal environmental impact" (U.S. Congress 2009, 3–5). The document also notes that "the biggest environmental footprint for electronics arises out of their production" (U.S. Congress 2009, 4). Meanwhile, in a section titled "Background," the matters of concern shift to issues of governance and export. Governance is seen to be a problem because of the patchwork landscape of state-legislated electronics take-back and recycling programs in the United States, in contrast to the regional approach of the EU (for an analysis of the U.S. legislation, see Lepawsky 2012). "Another significant problem," runs the report, "is the export of e-waste from the developed world to China and other developing nations" (U.S. Congress 2009, 5). In support of this claim, the text of the hearing states, "According to the Basel Action Network (BAN), approximately 80 percent of the e-waste directed to recycling in the US is not recycled, but is instead exported" (U.S. Congress 2009, 5). A few pages later, a prepared statement from one of the congressional committee members, Representative Eddie Bernice Johnson, appears. She states, "It really is stunning to know that the United States exports 80 percent of its electronic waste" (U.S. Congress 2009, 12). No representative of BAN appeared at the hearing, and the organization provided no written testimony that appears in the hearing documents. Despite a list of post-hearing questions that makes up a ten-page appendix, no one asked any questions about the hearing's background information on exports, though questions were asked about other aspects of the hearing's background topics, such as legislation. The expert testimony of the hearings accepts this figure of 80 percent exported as a known fact. With just a few pages of text separating them, a piece of "background" information attributed to a specific source (BAN) becomes a claim to knowledge detached from it. As a statement about the scale of exports, the 80 percent figure floats free of attribution, untethered from the circumstances of its production. How is it that this particular claim to knowledge has been able to float free of its source? To become free of any need to test the foundations on which is built? How did it become settled as a matter of fact?

To answer these questions and explore their import, I make use of several moves derived from different patches of the science and technology studies (STS) literature. One of those patches is what the feminist science studies scholar Myra Hird calls the problem of waste's indeterminacy (Hird 2012, 454). The problem, Hird argues, citing Brian Wynne, is that "despite their ubiquity, wastes exist in a twilight zone where no clear, 'natural' definition of them can be given, within wide margins of uncertainty" (Hird 2012, 454; see also Wynne 1987, 1). The contrasting examples above of millions of pounds of e-waste exported and the vastly larger volume of waste acid arising from smelting are part of the point Hird and Wynne are making. But there is an even more fundamental way in which the inherent indeterminacy of waste matters for how we attempt to know it. No matter how we measure waste, indeterminacy creeps into the core of our understanding. E-waste is categorically indeterminate. Even if we want to measure it, corralling "it" into a coherent category of the waste stream for analysis generates rather than eliminates uncertainty. Here is why: electronic items in a host of forms are ubiquitous but unevenly distributed things. They have been, and are being, incorporated into a vast range of other things that one might not intuitively associate with electronic waste. For example, in 1977 the auto manufacturer Oldsmobile put a microcomputer in a production vehicle for the first time (Bereisa 1983). Since then the amount of software in premium model cars has expanded to millions of lines of code (Broy et al. 2007; Charette 2009). Broy and co-workers (2007, 356) contend, "The next generation of upper class vehicles, hitting the market in about five years [ca. 2012], is expected to run up to 1 GB of software. This is comparable to what a typical desktop computer runs today [ca. 2007]." Vehicle manufacturers have attempted to use the U.S. Digital Millennium Copyright Act to prohibit owners from repairing their own vehicles because of the proprietary software running key engine components. So what are such vehicles for the purposes of measurement? Are they computers and—when they later enter a waste stream—e-waste? This is not a trivial question.

While there is a great deal of contemporary publicly expressed concern over "e-waste," especially over transboundary shipments of it, there is no comparable expression of concern over "c-waste" (car waste), at least not without substantially changing the terms of debate to be about such issues as vehicle emissions, air pollution, and climate change. Compared with the current public profile of concerns about where the bodies of discarded electronics end up, much less is said now about where end-of-life vehicles go. Historically, the situation was quite different.

In her classic work, *Waste and Want*, Susan Strasser (1999, 192) argues, "Automobiles provide the most prominent illustration of the workings of obsolescence in the decades before the Depression." What Strasser traces is a decades-long process in which manufacturers and consumers engaged in a different dance of commerce, one no longer premised on the durability of commodities but instead on the possibilities of their mutable fashionability—and the disposability that afforded. This was learned behavior, sometimes strongly resisted but taught initially through clothing designs marketed to middle-class women of late nineteenth-century America. The disposability of the new class of ready-made garments and what was understood as their wider influence on the design of other household items were being lamented by critics as early as 1867 (Strasser 1999, 189). What Strasser's work shows so well is that certain classes of discarded objects can become allegorical emblems of broader social changes—lamented by some, lauded by others—but they are not preordained as such. Waste never just *is*.

Automobile manufacturers learned well from the lesson of fashionability that "markets were not shaped by preexisting supply and demand but could be developed and extended" if, for example, they offered a range of models with variable options at different price points (Strasser 1999, 188). As automakers turned to fashionability to drive sales, turnover increased. The nexus of fashionability and disposability in the still new auto industry spawned reuse markets for used cars where none had previously existed, but along with it came a new geography of waste in the form of abandoned vehicles. In the forty-year period between 1920 and 1960, "automobile graveyards" became a prominent national concern in the United States among the growing membership of automobile associations and, after World War II, among the rapidly expanding suburban populations of America (Zimring 2011, 528). As early as 1922 the Automobile Association of America was expressing concern about cars abandoned on the nation's roadways as an unacceptable aesthetic blight. By 1935 Walker Evans, the legendary Depression-era photographer, was picturing piles of abandoned automobiles as iconic images of the crisis (Strasser 1999, 160). The postwar boom in mass production and consumption generated a situation in which Americans were buying five million new vehicles annually while scrapping up to eight million per year (Zimring 2011, 527). By the 1950s furtive heaps of cars had become visible "amid the shopping malls and residential subdivisions springing up around them" (Zimring 2011, 529) and had come into conflict with burgeoning suburban aesthetic sensibilities. So too did the automobile scrapyards, as did the often racialized and classed owners and workers who handled those growing masses of abandoned vehicles—a linking of discards, identity, and space that puts the lie to any notion that waste is ever merely a neutral technical designation of materials no longer wanted or useful. (On the constitutive links among waste, space, and identity formation, particularly race and class, see, e.g., Bullard 1994; Dillon 2014; Zimring 2005, 2016). As late as the 1970s, images of automobile graveyards still had enough symbolic value to serve as emblems of a newer American economic malaise and a growing ecological crisis (see figure 5.2) (see also Minter 2013b; Simmons 2009. For a historical treatment of auto salvage in the United States, see Lucsko 2016; on the environmental effects of automobiles throughout their life cycle, see McCarthy 2007; and see Brooks 2012 for an analysis of the international used car trade).

By the mid-1970s, however, the discourse around the abandoned car problem had changed, thanks to a new technological fix: the automobile shredder. Shredders, or "car-eaters," which could "digest 1,400 cars a day," were reported to be "the most practical answer to the disposal problem" (Carr 1969, cited in Zimring 2011, 534). By 1975, with the widespread use of the shredder, the automobile graveyard problem was, ostensibly, over. But, in a nice parallel with themes I pick up in the next chapter, one way of managing a given waste problem transformed it into another waste problem rather than actually getting rid of it. Auto shredders did mitigate the problem of discarded vehicles as an issue of aesthetic blight. They also vastly improved the economics of scrap recovery by radically reducing the labor time necessary to separate automobiles into their constitutive materials. Yet it was exactly this technological change that transformed the problem of auto waste from a largely aesthetic concern into a chemical

Figure 5.2
Abandoned automobiles near Great Salt Lake, 1974.
Source: National Archives ARC 555849, photograph no. 412-DA-13397, Documerica
Series.

one. The basic workings of auto shredders generate both ferrous and non-
ferrous scrap, but also what the industry calls "fluff," or automobile shredder residue (ASR). The toxic hazards of ASR result from the heterogeneity
of automobiles themselves, especially as a consequence of the plastics and
polyurethane foams increasingly being used in their construction. Shredders were producing 2.5–3 million tons of ASR annually in the United
States by the 1990s (Schmitt 1990, cited in Zimring 2011).

The year 1965 saw the passage of the Highway Beautification Act in the
United States. Among other things, the act mandated that junkyards along
interstates and highways "be removed or screened" to encourage "scenic
enhancement and roadside development" (Federal Highway Administration, U.S. Department of Transportation 2017). This was a legislative
response to automobile waste as a problem of aesthetics. The year 1970
saw the creation of the U.S. Environmental Protection Agency (EPA) with
jurisdiction over the control of the effects of hazardous materials and
hazardous waste on the environment. During the 1990s, "dozens of scrapyards were designated as Superfund sites by the EPA," partly in response

to the release of ASR into air, soil, and water (Zimring 2011, 542). Aesthetic blight had become toxic hazard.

The shifting geohistorical ground about what counts as a waste crisis and what such crises are parables for hints at how given objects or materials deemed emblematic of that crisis are loaded with allegorical meanings, as opposed to merely being extant problems of technical waste management. As Wynne shows, when it comes to waste (particularly hazardous wastes), "natural processes and human interactions are jumbled together in complex and widely variable ways, making a badly structured and, indeed, indeterminate behavioral-technical-risk-generating system" (Wynne 1987, 1). The contrast in relative contemporary concern over e-waste versus anything like c-waste is useful because it emphasizes how e-waste is as much a problem of knowledge—epistemology—as it is about finding workable solutions to a given waste management problem. The contrast also points to the role of allegory in problems of knowledge. In attempting to know e-waste, statistically or otherwise, we are also telling stories with moral lessons built into them.

No knowledge about e-waste, whether it arises from quantitative measures, documentary photography, or any other mode, is morally neutral. But pointing out that all knowledge comes with allegorical meanings is not the same thing as claiming that all knowledge does so in the same ways or with the same effects. How we know matters. The indeterminacy of knowing e-waste does not end even if shared agreement about what things constitute it can be arrived at. For example, the majority of regulatory frameworks for managing e-waste are premised on metrics of weight. But weight tells us nothing about toxicity. So even if we were able to generate precise knowledge about e-waste in terms of weight, that would do little to tell us about its propensity for ecological harm. No nonarbitrary criteria exist that will adjudicate between measures of weight and measures of toxicity to tell us which is the best or right measure to use. To choose one or the other mode of measuring is political in the sense of making decisions with potentially transformative consequences in a terrain of uncertainty. Measuring is unavoidably political in this sense.

How we measure matters immensely since different measures suggest different ways of dealing with e-waste when construed as a problem of how to manage it. For example, the financing models of legislated take-back systems began running into financial difficulty after only a few years in operation precisely because of the way their financing formulas depended on measures of weight. Since these systems have been put in place, the average weight of manufactured electronic devices has fallen faster than the total mass of devices collected. Consequently, an increasingly perverse situation exists in which more and more increasingly lighter devices are being collected yet, because the fees collected to handle them are typically premised on weight, the fees collected are falling below the costs of recycling the products. The situation is leading to bankruptcies of recycling firms and large inventories of abandoned discarded electronics (see, e.g., Washington Materials Management and Financing Authority 2015, 38).

As I discussed in chapter 2, Wynne (1987) argues that the desire for ever greater precision in definitions and measures of wastes, particularly hazardous waste, is actually counterproductive to their effective management. The counterproductivity of the desire for precision arises because it is impossible even in principle to categorically define either waste or hazard or risk; each is fundamentally indeterminate (see also Gille 2013; Hird 2012, 2013). Knowledge about materials deemed to be hazardous waste is always incomplete, and there are no physical properties of materials that can lead to their fixed and universal designation as risky hazardous waste.

Yet indeterminacy is no excuse for inaction. Deniers of global warming science have used the notion of indeterminacy to argue in favor of business as usual. We must wait until the science is certain before acting, so the claim goes. The horrible irony, of course, is that certainty will arise only once the full effects of global warming are upon us, by which point it will be too late to do anything about it (see Oreskes and Conway 2010 for a detailed investigation of the tactics of climate change deniers). Efforts to use science to justify policy making around hazardous waste face a similar conundrum that the search for precision will never solve. As Wynne (1987, 8–16) points out, the more that scientific expertise is used to justify policy decisions, the more scientific practices come under scrutiny by variable and disparate publics. When the inner workings of scientific practice fail to match up to widely shared but totally unrealistic images of how scientific practices work, the credibility of science and of the regulatory process is eroded. Yet such claims about the need for scientific certainty before action is taken persist only because of a pervasive story—one is tempted to say myth—about what science does, namely, that it gives us proofs. Yet as the vast STS literature shows, the sciences do no such thing—save mathematics, where "proof" has a particular meaning quite distinct from its commonsense definition (for foundational studies, see Latour 1987; Shapin and Schaffer 1987; for accessible introductions to these ideas, see Caley 2007; Lewis 2014). So by highlighting the inherent indeterminacies of waste and its measurement, I am not also making a claim that action to mitigate e-waste should be deferred until we have certainty about it. I am concerned with quite a different problem: what is the right thing to do with e-waste *despite* the inherent uncertainties that come with any attempt to know it.

Instead of proof, science and technology give us more or less trustworthy knowledge and infrastructure. However, debates about e-waste typically remain framed as a battle between truth and falsity. Yet when e-waste debates are framed in such binary terms, they quickly enter the same cul-de-sac as the spurious debate between climate scientists and climate change skeptics. That cul-de-sac is built on a strangely shared ground between the two sides and their phantom publics: an opposition between science and politics. In this schema, science is supposed to be what grounds facts as distinct from values and, again supposedly, it is science that provides the neutral and objective grounds for nonpartisan decision making. Politics, on the other hand, is presumed to be rife with arbitrary value judgments and is understood to be only what corrupts and pollutes the facts that science gives us. This is an unsustainable politics of knowledge; it is a dead-end street that crumbles because of its own poorly mixed materials (Latour 1993, 2015). While this may seem to be an abstract philosophical point about how we know what we know (epistemology), its implications are concrete: there is no neutral set of criteria by which we can categorize or measure e-waste. We may choose to measure weight, but then we miss toxicity. We may measure toxicity, but, as the text of the Basel Convention notes, the "potential hazards posed by certain types of wastes are not yet fully documented; tests to define quantitatively these hazards do not exist" (Basel Secretariat 2011a, 59) (see chapters 2 and 3 for a detailed discussion of the Basel Convention). Or we may define e-waste as a problem of post-consumer discards, but then we miss all waste from the upstream practices that mine, design, and manufacture electronics and give a material fate to that which will eventually be discarded.

Noting that scientific practices result in more or less trustworthy knowledge (as opposed to truth, falsity, or proof) points to why it is important to pay attention to the work of methods. Another patch of STS literature, sometimes referred to as research on the social life of methods (Law 2004a, 2010; see also Barad 2007), helps me explore the significance of the methods used to know e-waste. This literature argues that methods cannot be understood as neutral tools that merely observe the world. John Law (2004a, 2012) has written extensively on what he calls the double social life of methods. Methods are doubly social in that they are situated (i.e., they are articulated by and from particular places and people) and they have effects (i.e., they are epistemologically and ontologically generative). The effects of methods enact a given phenomenon in particular ways and not others (e.g., e-waste as a postconsumer waste management problem as opposed to an issue of occupational health and safety on the manufacturing line). By enacting *this* way and not *that* way, methods partially make the worlds they usually purport only to study.

If methods partially enact the worlds they investigate, then one way they do so is by making some things present and others absent so as to make sense of the worlds methods enact. The necessity of methodological presencing and absencing is not as abstract as it may seem. Imagine trying to make sense of the world with some sort of panoptic camera that could literally take a picture of *everything* in one frame, including the camera itself. How would a viewer of that picture have any idea what the researcher is providing such a photograph as evidentiary material of since *nothing* is excluded and everything is present? Another way of thinking about this same idea is through the analogy of the map famously narrated by Jorge Luis Borges in his single-paragraph short story, "On Exactitude in Science" (see also Self 2013). In it Borges describes an empire in which the science of cartography is so advanced in accuracy and precision that only a map at a one-to-one ratio is deemed adequate by the cartography guilds and the rulers of the Empire. Yet as Borges's story unfolds, such a map falls out of favor with subsequent generations. It is not hard to understand why: a perfectly matched map and territory make both map and territory perfectly useless. Finding one's way on the map is exactly as difficult as finding one's way on the terrain.

Some things *must* be left out in order to be useful for analysis, interpretation, and explanation. This necessity points to the inherent politics and ethics built into any research strategy. If by "objective" one means to point to methods that are neutral, then no method can achieve that idealization of how research works. Decisions about what to count, what not to count, and how to count are, by definition, decisions made in a terrain of uncertainty (if certainty were present, there would be no need for research). There are no nonarbitrary criteria by which to make those decisions. In that sense, methods, like measures, are political and raise issues of ethics. All methods have an ethic to them—claims about good and right action—if in no other way than in the sense of what should be made present and what should not. To imagine that a given method merely *is*, neutrally at rest between the world and the methodologist, ignores the *oughts* inherently built into any method. Answering the question of what to make present and what to make absent is a performance of such ethics: it is good and right to make X present and Y absent.

The action of methodological presencing and absencing generates what Law calls collateral realities. These are "realities that get done incidentally, and along the way" when any knowledge-making practice—any method—gets done (Law 2012, 156). While collateral realities may seem like an abstract philosophical construct, the concept is grounded in an aspiration to a certain kind of realism in research. The moment a researcher decides she wants to study e-waste and poses questions such as *How much e-waste is produced? Where? By whom? Where does it go?* and presupposes e-waste as a problem of postconsumer discard, then collateral realities are already being generated even before particular methodological tools (e.g., interviews, surveys, trade data, photography) are put to work. Her questions enact e-waste as a postconsumer discard rather than any other way of enacting it (e.g., as a phenomenon of resource extraction, or groundwater contamination from manufacturing, or occupational health and safety on assembly lines). Since the questions posed partially enact the thing to be researched—here, e-waste—in particular ways and not others, the reality to be researched could be otherwise. In this sense, because different questions and methods enact different e-waste realities, the collateral realities arising in their wake are "through and through political" (Law 2012, 156).

Exporting Harm as Theater of Proof

As Law argues, the enactment of collateral realities is typically unintentional. Such enactments may or may not be good. Indeed, they may be obnoxious, even pernicious. In 2002 BAN published a report titled *Exporting Harm: The High-Tech Trashing of Asia* (Basel Action Network 2002). The

first page of that report states, "Informed recycling industry sources estimate that between 50 to 80 percent of the E-waste collected for recycling
in the western US are not recycled domestically." Several different versions
of this statement appear elsewhere in the report. For example, on page 11
the report reads, "Informed industry insiders have indicated that around
80% of what comes through their doors will be exported to Asia, and 90%
of that has been destined for China." A few pages later, under the heading
"How much E-waste is Exported?," the report frankly states, "Nobody really
knows" (Basel Action Network 2002, 14). But then it continues, stating,
"Anecdotal evidence on E-waste exported by the US to Asia is abundant"
and "Very knowledgeable and informed industry sources, however, have
estimated that around 80% of what is diverted to recycling [in the United
States] is actually exported to Asia" (Basel Action Network 2002, 14). An
endnote following the latter claim attributes it to a single telephone interview BAN conducted with someone named Mike Magliaro of Life-Cycle
Business Partners in Salem, New Hampshire, in February 2002 (Basel Action
Network 2002, 14, 50 endnote 41).

The notes of equivocation in *Exporting Harm* are important to keep in
mind. Certainly the BAN report makes the case for the export of e-waste
in terms of crisis. But it does also provide a range for the export problem
(50–80 percent). And it does explicitly concede uncertainty about the
actual scale of the problem ("nobody really knows"). The indeterminacy
apparent in *Exporting Harm* disappears, however, once the 80 percent figure
enters the fray of the 2009 congressional hearing for the Electronic Waste
Research and Development Act of 2009, where it is cited as fact. What
accounts for the transition from more hesitant claim to settled fact? What
is it about the BAN report that makes its claims so compelling? Why has
it become, in Latour's words, such an effective "theater of proof" (see
Latour 1988, 85–86, 1999, chap. 4)?

Part of the answer lies in how *Exporting Harm* arranges its claims, evidence, reasons, and warrants. The invocation of numbers in the report
relies on modes of representing those numbers in ways that bolster their
claims to trustworthy knowledge. For example, *Exporting Harm* attributes
the figure of 50–80 percent to "informed industry insiders" and "very
knowledgeable and informed industry sources," yet does not state who the
sources are except in one instance (the phone interview with Mike
Magliaro). It is a move that suggests there are many experts that make the

same estimate about e-waste exports. This pluralization of support is important since it would seem to bolster BAN's claims.

The use of deference to other, external authorities is part of a literary technique that STS scholars sometimes call "modest witnessing" (for a foundational study, see Shapin and Schaffer 1987; for trenchant critiques of the assumptions built into modest witnessing, see Haraway 1991b, 1997). In short, modest witnessing is a literary technique in which the authors of a truth claim present that claim in such a way as to suggest that the claim would be judged true by anyone who followed the same methods as the author and, more important, that the claim so judged does not come from the authors but from some *other* authoritative sources such as instruments or acknowledged experts. Need it be said, modest witnessing is a representational tactic that is neither unique to, nor scandalous to find at work in, BAN's report (it is, after all, clearly at work in the congressional hearing, as well as in this very book). But the pluralization of support in *Exporting Harm* from one named source to an untraceable number of other sources is a move that implicitly asks readers for their trust. Given the influence of *Exporting Harm*, it is important to ask, then, why it is deemed trustworthy? What else about the report might have led to such widespread credibility?

Some of the answers are found in how the text of *Exporting Harm* uses other forms of evidence, particularly the visual evidence of documentary photography. The geographer Gillian Rose (2001, 20) writes of documentary photography that it "originally tended to picture poor, oppressed or marginalized individuals, often as part of reformist projects to show the horror of their lives and thus inspire change. The aim was to be as objective and accurate as possible in these depictions." There is, of course, a long tradition of critical analysis that questions such underlying assumptions of objectivity and accuracy in documentary photography and film (e.g., Rosler 2003; Sekula 1978; see also Demos 2013). One of the reasons why documentary methods such as photography and film are so powerful—that is, why they come to circulate, inform, and actively shape the discourse of which they are part—is that they are able to draw on traditions of witnessing, especially first-person or eyewitnessing, for their authority to claim trustworthiness about that which they are claiming to have knowledge. Documentary photography generates what Schwartz and Ryan (2003, 8) call "photographic seeing," that is, "a surrogate for first-hand observation." In this sense, the conventions of documentary photography bolster claims premised on the technique of modest witnessing. Yet even on its own terms documentary photography is inevitably selective. It cannot avoid being so, since no picture can show everything. Thus, understanding the selectivity of such images means paying closer attention to how they are generated in the first place. Yet absent from *Exporting Harm* is any publicly disclosed sampling frame used to obtain the images included in the text. Without such a frame it is impossible to know to what extent a given image is a singular instance or part of a general pattern.

The indeterminacy generated by the documentary photographic practices in *Exporting Harm* extends to other visual evidence provided in the report, such as images of asset tags. Asset tags are labels sometimes affixed to electronic equipment, particularly by organizational users such as large businesses, government institutions, and the like. Such tags help organizations track equipment assets when they are in their inventorial control. Images of such tags appear in *Exporting Harm*. For example, on page 14 of the report a photograph appears with asset tags from the City of Los Angeles, a California medical facility, and the Los Angeles Unified School District. The caption of the photo reads: "Just some of the many institutional labels from the United States found on computers in Guiyu, China in, December 2001" (Basel Action Network 2002, 14). There is no doubting the tags themselves. There they are in the image, after all. Yet in the absence of any publicly disclosed sampling method the tags (and the image of them) tell us very little. How many are "some"? How much is "many"? Compared to what? Were the tags selected out of a larger collection to tell a particular story about exports? They also tell us nothing about how long, if at all, the equipment may have been in a direct reuse phase after export, which is important since the Basel Convention explicitly allows exports for direct reuse (see chapter 2). A reader of the BAN report has no way of knowing how representative these tags are. Yet as photographs, the images can draw on the hinterlands of photographic seeing to lend credence to the claims being made in the text of the report.

Similar points can be made about the use of forensic data analysis of memory devices. This approach involves recovering data recorded on devices such as hard drives to determine previous ownership. It is a technique used to compelling effect in Peter Klein's (2009) Emmy Award–

winning investigative journalism piece, "Ghana: Digital Dumping Ground," for PBS's *Frontline*. In that piece, journalists using hidden cameras are shown purchasing hard drives from a dealer near Agbogbloshie, an infamous "hot spot" of e-waste dumping. The drives are then taken to a nearby university where a computer scientist downloads data available on the drives. The retrieved data include photographs, banking information, and other files of a personally identifiable character. On another drive, data from U.S. government contracting bids are found, including information from the Northrop Grumman company. Much like the photographic evidence discussed above, forensic data analysis offers a compelling sense of witnessing the truth. Yet, while there is no reason to doubt the specificity of results obtained from the drives, the knowledge forensic data analysis can provide about the e-waste problem does not escape indeterminacy. The *Frontline* documentary does not use any sampling method, so viewers have no way of knowing how representative the particular drives are of origins. The documentarians also purchased the drives from a business selling them for reuse—again, something the Basel Convention expressly permits. Devising a sampling strategy that could account for systematic biases between reuse markets and waste markets is extremely difficult if possible at all. For example, depending on where in the trajectory of the "afterlife" of electronic equipment one can or cares to look, one would need to account for equipment having been sorted for reusability or high material value. Neither of those variables is either static or universal. The journalists producing the documentary account for none of these issues. Moreover, sampling-related issues like these are only part of the indeterminacy generated by forensic data analysis. Relying on this approach limits knowing e-waste to devices that contain memory storage components. Many categories of devices, however, do not contain such components (e.g., monitors and input peripherals such as keyboards and mice), which limits the utility of forensic data analysis for making claims about e-waste as a general category of the waste stream.

Some Collateral Realities of Exporting Harm

Tracing the effects of the knowledge-making practices of which *Exporting Harm* is exemplary is useful, but not because doing so bolsters a critic's ability to impugn the ENGO as a purveyor of false, made-up, or fake numbers. What is notable about *Exporting Harm* is not the unreality of its numbers but quite the opposite: the real and substantive work those numbers have achieved. Instead, following the traces of *Exporting Harm* is useful for two quite different reasons. First, doing so helps illuminate the social lives of methods used to enact e-waste as an object of knowledge, concern, and management. Second, tracing the effects of *Exporting Harm* is useful for charting a social formation in the making, a collection of actors—including myself—that condenses in the wake of the report's circulation and is grouped by our shared interest in e-waste as a matter of concern. To follow the trace of the report is to hint at how following such traces can open onto a broader project of mapping the publics and their politics sparked into being when e-waste becomes a public issue or matter of concern. In what follows I pay particular attention to the publics and politics formed in two networks in which *Exporting Harm* circulates: academic literature that cites the report and public websites on which the cover image of the report appears.

Some of the collateral realities of *Exporting Harm* are traceable using techniques such as citation analysis (for nontechnical introductions to these techniques, see Archambault and Larivière 2010; Bornman 2013). Using Scopus—a bibliographic database familiar to many academics across the fields of social sciences, humanities, sciences, technology, and medicine—one can easily trace how *Exporting Harm* has been cited. At the time of writing, Scopus held over 2,800 documents with the search terms "e-waste" or "electronic waste" in their title, abstract, or keywords. With over 270 citations, *Exporting Harm* is the sixth most cited document in this corpus of literature. If the bibliographic search is limited to articles (i.e., excluding conference papers, notes, book chapters, and the like), *Exporting Harm* is the second most cited document in the literature. This citation pedigree is notable for several reasons. First, *Exporting Harm* is not a peer-reviewed publication, yet it is cited far more frequently than the vast majority of peer-reviewed literature on the topic of e-waste. Moreover, Scopus citation data for the report show an increasing trend in citations to it over time. It is most frequently cited in the fields Scopus classifies as environmental science, and indeed, it is most frequently cited in the journal *Environmental Science and Technology*. These broad citation trends are indicative of what others are doing with the *Exporting Harm* report. In that sense, they are traces of the report's collateral realities, that is, realities brought into being, not necessarily intentionally, along the way with the report's use by others.

More detailed outlines of these collateral realities can be surveyed by turning the titles, abstracts, and keywords of the literature citing *Exporting Harm* into a corpus of text organized by year of publication. This body of work can then be charted using techniques for textual analysis, much as a cartographer surveys an unknown terrain with surveying instruments (for an introduction to textual analysis, see Rockwell and Sinclair 2016; for a specific guide to the tools used to generate the textual analysis I discuss here, see Sinclair and Rockwell 2016).

Initial readings of the topography of these texts point to consequential results. The corpus comprised of the titles, abstracts, and keywords of the more than 270 documents citing *Exporting Harm* amounts to over 65,000 words—about the length of this book. The five most frequent words in the corpus, ranked from highest to lowest occurrence, are "waste," "environmental," "recycling," "electronic," and "management" (see figure 5.3). This finding hints at the shape of the scholarly conversation defined by those documents citing *Exporting Harm*. In that conversation, recycling predominates as a strategy for managing electronic waste as an environmental concern. Yet we know that postconsumer recycling cannot solve waste problems arising upstream in, for example, raw material extraction, manufacturing, or distribution. *Exporting Harm* explicitly acknowledges this latter point. The "solution [to e-waste] lies upstream," notes the BAN report (Basel Action Network 2002, 40). "Pollution prevention," the report rightly argues, "does not just mean recycling waste already produced— rather it means clean production—producing less quantity of waste and less hazardous waste in the first place" (Basel Action Network 2002, 40). Yet the realities of e-waste enacted by the scholarly literature citing the BAN report pay almost no attention to the role of upstream solutions to e-waste. None of the titles, abstracts, or keywords of the documents citing *Exporting Harm* makes even one reference to "clean production." A search through the corpus for any word variants of "design," "manufacturing," "reuse," "repair," or "producer responsibility" suggests anemic attention at best to their role in the generation of what will, eventually, become postconsumer e-waste (see figure 5.4). Variants of "recycling" occur 439 times. Indeed, references to variants of "recycling" predominate across all years of publication (see figure 5.5 online). Contrary to the importance of upstream solutions professed in *Exporting Harm*, it is downstream recycling that is the dominant framing of management solutions to e-waste as a problem in the scholarly literature citing BAN's report.

Figure 5.3
The most frequently occurring words in titles, abstracts, and keywords of documents citing *Exporting Harm.*

		Term	Count	Trend
☐	1	recy*	439	
☐	2	design*	49	
☐	3	manufac*	47	
☐	4	"producer responsibility"	39	
☐	5	reus*	38	
☐	6	repair*	5	
☐	7	"clean production"	0	

Voyant Tools . Stéfan Sinclair & Geoffrey Rockwell (© 2017) Privacy v. 2.3 (M1)

Figure 5.4
Counts of the appearance of key terms in the corpus of literature citing *Exporting Harm.*

Scholars have themselves largely worlded e-waste as a problem of post-consumer discard management. As such, recycling becomes the overwhelmingly dominant way to imagine solutions to the problem in academic literature that cites *Exporting Harm*. A scholarly public around e-waste has gelled and formats the issue in a very particular shape, one predominantly configured by recycling. Yet the very report this academic literature cites argues that producing less waste in the first place is more important than recycling to the overall goals of pollution prevention. Here, then, a collateral reality results from the circulation of BAN's report into scholarly citation networks: a conversation among academics devoted almost entirely to recycling and one that says almost nothing about any of the most critical activities to avoiding waste before it is made.

* * *

The wake of *Exporting Harm* can also be traced outside the secluded realm of scholarly bibliographies and through various spheres of the web. This is slightly more difficult to do since the web lacks the well-structured citational data sets curated for academic literature searchers. But a well-grounded set of techniques and tools has been developed by scholars interested in using the Internet to trace social formation and change not so much *on* the web but *with* the web (for foundational work in the field, see Rogers 2013; for an example of a practical implementation, see Rogers, Sánchez-Querubín, and Kil 2015).

One way to map the circulation of BAN's report on the web is to trace the circulation of images in the report and attributed to BAN. I performed such a search, tracing the recurrence of the photograph that appears on the cover of *Exporting Harm*. This is, of course, a selective and partial approach to tracing the work performed by *Exporting Harm*. Yet the results are telling.

The cover image of *Exporting Harm* depicts a child, presumably a boy, with Asian features. His bare feet are smudged black with dirt and/or the residues from the pile of trash he sits atop. He is wearing a sweater with bright yellow and green patterning. In the background are tall green grasses and behind them is a clutch of small two-story buildings. The boy looks directly into the lens of the camera and thus into a viewer's own eye. The photograph has no caption in the report, but it is available on BAN's Flickr page. There the image's caption reads, "Migrant child from Hunan province sits atop one of countless piles of unrecyclable computer waste imported from around the world" (Basel Action Network 2016b). Many scholars have written critically about the use of images like "Migrant child" for how they ostensibly reinforce the financial churn of various NGO and donor industries (e.g., of "development" or "the environment"). Such images are accused of aestheticizing the horrors—even the pornography— of poverty and suffering (e.g., Demos 2013; Jensen 2014; Ong 2015; see also Rosler 2003; Sekula 1978). Critiques of this sort are important. But as Demos (2013) argues, to accuse such images, their makers, and their viewers of "'aestheticizing horror' is too convenient" because doing so ignores "the complex entanglement between the aesthetic intensity of the exceptional situation taken in by a gaze, and the ethical or political concern to bear witness to the horror of a reality nobody is bothering to see" (Demos 2013, 114–115, citing Rancière). The solution to such spectacular depictions of misery is "not to look away … but rather to look again in a different way" (Demos 2013, 114). So, rather than look away from "Migrant child," I will take Demos's suggestion and look at it differently by surveying the sites and situations where it appears beyond the report for which it provides the cover image. In so doing, I largely forgo interpretation of the image content and instead examine the textual content of the sites where the image appears online.

To do so I use a combination of techniques and tools, including appropriating Google's reverse image search capabilities for this purpose (for an overview of the procedural considerations for this kind of research, see Digital Methods Initiative 2010). Google's image search platform enables a user to drag and drop a photo into Google's search bar. Through a process of pattern matching combined with the company's proprietary search algorithm, Google then provides a "best guess" for the image in question (Google 2011). It also helpfully provides a collection of "visually similar images," followed by links to pages that include a match of the image searched. When I performed such a search on May 30, 2016, using the cover photo of *Exporting Harm*, Google's reverse image search showed me there were about 392 results, and that the "best guess for this image" was "e waste developing countries" (see figure 5.6). Before moving further into the analysis, I want to stop and consider the implications of this result because it speaks powerfully about what Law means by the unintentional enactment of collateral realities and what Pine and Liboiron (2015) refer to as charismatic data, or data with sufficient allure to "move an audience to action because of how the data resonates with preexisting cultural values, desires, and morals" (Liboiron 2015, 16).

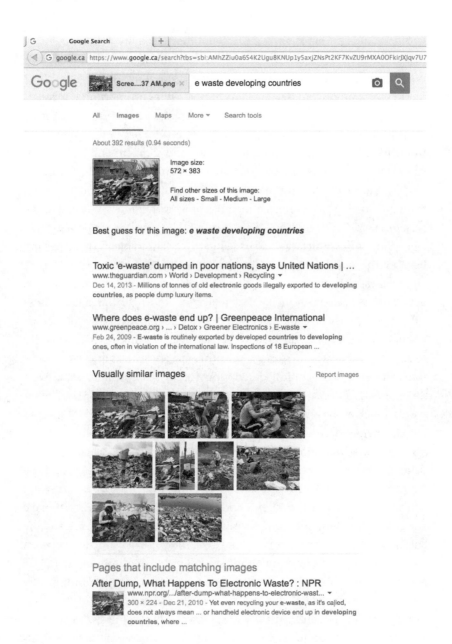

Figure 5.6
A screenshot of a Google reverse image search conducted May 30, 2016, of the cover photo of BAN's *Exporting Harm* report.
Source: Google and the Google logo are registered trademarks of Google Inc. Used with permission.

I have already made the point about the selectivity of documentary images as a method for knowing e-waste and the indeterminacy such images generate in the absence of systematic sampling. Here, where cover photo meets proprietary search algorithm, outlines of a portfolio of collateral realities, both determinate and indeterminate, can be traced. What I want to draw attention to is the specificity of the image (a migrant child from Hunan; unrecyclable computer waste, and so forth) and how it contrasts with the results of Google's pattern-matching search algorithms. A specific image taken on a specific date at a specific location becomes, through Google's algorithms, a member of a vastly more general class: "e waste developing countries." This does not happen by magic. Google's algorithms are proprietary and consequently unknowable in their specifics beyond the small group of humans with proprietary access to them. However, search engines, including Google's, work by returning results generated by general principles of relevance known to outsiders. These principles of relevance are determined by a calculative mixture of factors, including the number of links a URL receives ("inlinks"), its popularity (e.g., how many users click on it), and other factors, such as its currency and longevity (for nontechnical descriptions of these principles, see Rogers 2013, 30–31; Rogers, Sánchez-Querubín, and Kil 2015, 34–35). In other words, much as the results of a Google search (or that performed by another search engine) are determined by algorithms, the performance of those algorithms in returning results deemed relevant to the search engine's human users is partly dependent on the collective, tracked action of those human users (Feuz, Fuller, and Stalder 2011). This is why Rogers (2013) and others (e.g., Marres 2012; Marres and Gerlitz 2015; Venturini 2010) make the case that the Internet can be turned into an instrument for researching social formation and change (they make the conceptual distinction between researching the social *with* the Internet as opposed to *on* the Internet).

Out of the collective action of search behavior—the algorithmically equipped social-in-the-making—comes the collateral reality of "Migrant child from Hunan" enacted as "e waste developing countries." A substantial jump from a specificity to a generality that would otherwise be unattainable by a single image on its own is rendered into existence. In a profound sense, the jump in scale is a not necessarily intentional result achieved when knowledge made in the absence of systematic sampling methods (e.g., a specific documentary photograph) meets the systematicity of search algorithms and the human search behaviors that reinforce the results deemed relevant to those users. It is telling, for example, what sorts of results are *not* returned by the image search even though, in principle, they could be. For example, the search using the "Migrant child" image does not produce results such as "buildings, low two-story," or "child wearing sweater," or "grasses, tall, green," or any other of the almost infinite possible ways of making sense of the "Migrant child" image. In effect, one is witnessing the consequences of image content resonating with the values, desires, and morals shared widely enough by users of the search platform to return the results of a shared understanding of e-waste in developing countries. This shift from specificity to generality is a profound enactment of a collateral reality of "Migrant child," pointing, as it does, to social formation in action. And the *action* of that formation can be glimpsed by comparing results for the same search done almost a year later, on April 26, 2017. This second search resulted in more than 25 billion pages containing the "Migrant child" image—and a more refined result for Google's "best guess." The image is now associated with "guiyu china e waste" (see figure 5.7).

From the reverse image search described above, I collected the first one hundred working, nonduplicative URLs on which "Migrant child" appeared. I then scraped all text from each of those one hundred URLs using a web-based scholarly text analysis tool called Voyant (for an introduction to Voyant, see Sinclair and Rockwell 2016). The text collected from each of these one hundred URLs amounted to a total of more than 125,000 words. Out of that total, the five most frequently occurring words were "waste," "green," "recycling," "electronics," and "China" (see figure 5.8). All variants of "recycling" appeared 823 times, compared to 94 times for all variants of "design," 89 for "reuse," 68 for "manufacturing," 19 for "repair," seven for "producer responsibility," and one for "clean production" (see figure 5.9). Much like the terrain of academic research on e-waste, the terrain of the web referencing *Exporting Harm* through the image of the migrant child is dominated by recycling. Like the scholarly literature citing the BAN report, the websites incorporating the report's cover image largely ignore any upstream solutions that would mitigate waste arising from electronics before it is produced.

"Migrant child" is a classic example of what the environmental humanities scholar Kevin DeLuca calls an "image event" (DeLuca 1999, 6). Such

Figure 5.7

A screenshot of a Google reverse image search performed April 26, 2017, of the cover photo of BAN's *Exporting Harm* report.

Source: Google and the Google logo are registered trademarks of Google Inc. Used with permission.

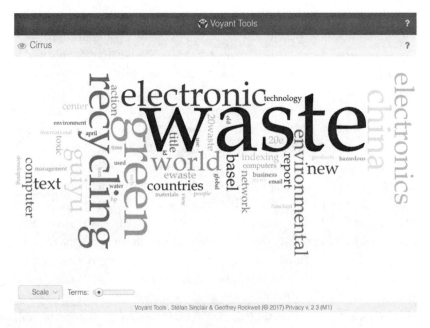

Figure 5.8
A visualization of the most frequently occurring words on websites that contain the "Migrant child" image.

Voyant Tools

Terms

	Term	Count	Trend
1	recyc*	823	
2	design*	94	
3	reus*	89	
4	manufact*	68	
5	repair*	19	
6	"producer responsibility"	7	
7	"clean production"	1	

? 7

Voyant Tools , Stéfan Sinclair & Geoffrey Rockwell (© 2017) Privacy v. 2.3 (M1)

Figure 5.9
The co-appearance of terms on websites that contain the "Migrant child" image.

events were pioneered by the environmental NGO Greenpeace, where BAN's founder, Jim Puckett, worked as toxics director before forming BAN (Basel Action Network 2016e). Image events are about gaining publicity for a cause, but they are also understood as weapons in the fight to change public consciousness. In the words of one of Greenpeace's founders, Robert Hunter, such images are "mindbombs" and mass communication technologies are the "delivery system" (Hunter 1971, 216). As DeLuca writes, image events are "crystallized philosophical fragments, mind bombs, that work to expand 'the universe of thinkable thoughts'" (DeLuca 1999, 6). *Exporting Harm* has undoubtedly crystallized e-waste as a matter of concern. Various publics have gelled in the shockwave of its detonation over a decade ago. Yet the universe of thinkable thoughts it has crystallized— evidenced in the scholarly literature and on the web—has a shape distressingly narrowed to a conversation mostly about what happens to electronics once they are discarded by consumers and recycled. I would claim, then, that the collateral realities of mind bombs entrain with them what other bombs do—collateral damage (see also Ingenthron 2011). As mind bombs, *Exporting Harm* and "Migrant child" have expanded the universe of thinkable thoughts about electronics to what happens to them once discarded. Yet, on the front of expanding that universe to solutions other than postconsumer recycling—such as moving solutions "upstream"—their traces through the social appear to have had little consequence. The publics represented by the academic citation of *Exporting Harm* and by the presence of "Migrant child" on websites have a very strong tendency to articulate e-waste as a matter of concern in terms of recycling. In the next chapter I attempt to show just how little postconsumer recycling will do to solve the e-waste problem when it is construed in terms of a postconsumer waste management problem for which recycling is proposed as the solution.

Conclusion

By engaging the techniques and tools for tracing the collateral realities of *Exporting Harm*, I have hinted at other ways of knowing e-waste. These tools and techniques are like any measurement instrument. Using them generates its own forms of indeterminacy about the knowledge built with them. For example, they make assumptions, such as that the frequency of occurrence of a word or phrase is an appropriate measure of its importance in a corpus of text. This is not an inherently neutral or objective assumption. A counterexample is a stop sign. It has one word on it. The significance or import of the word is its singularity, not its frequency. On the other hand, good scholarly practices for crafting article titles, abstracts, and keywords are supposed to condense the most salient aspects of a full article into their most distilled form. That is, titles, abstracts, and keywords are meant to bring the absolute essentials of an article to the fore for readers. Thus terms and phrases that appear in them represent what authors, editors, and reviewers have collectively determined to be the most relevant messages of a given article. Thus, to measure what is present and absent and with what frequency and compared to what seems a reasonable way to map the terrain of scholarly conceptualizations of e-waste as an issue.

With respect to using the Internet as an instrument of social research, Google is not the web, the web is not the Internet, and the Internet is by no means coterminous with "society" (Venturini 2010). Even with the vast computing resources of Google, its archive of the web can never be complete since the web is constantly being added to; search engine algorithms that crawl and archive the web will always lag the web's growth. And they typically overrepresent particular linguistic groups (especially the Anglophone worlds) as well as territorial jurisdictions, as studies of online access and censorship show (e.g., Graham and De Sabbata 2016). One can trace the social with the Internet, but doing so does not produce a one-to-one map of it.

More than fifteen years have passed since the publication of *Exporting Harm*. The efficacy of its knowledge-making practices is evident in its uptake in the expert testimony of the congressional hearing of 2009. Neither legislators nor witnesses ever questioned the trustworthiness of the 80 percent export figure. Indeed, it is the upper figure of BAN's own qualified range of 50–80 percent that makes it into the congressional record as background fact and known truth. Ultimately, Bill H.R. 1580 (111th) passed in the House, but died in the Senate and never became law (GovTrack 2009). It was revived a year later, only to fail again (GovTrack 2010), with yet another version introduced in February 2017 (Cook 2017). Meanwhile, *Exporting Harm* remains a widely cited source of estimates about e-waste exports. The figure of 50–80 percent exports has floated free of its original source and been taken up as known fact and even attributed to other sources, including my own work explicitly questioning the figure (see Wernink and Strahl 2015, 126).

In BAN's latest initiative using GPS trackers new numbers about export rates from the United States circulate. Now, however, the rates are said to be anywhere between 10 and 50 percent *below* the estimates from the *Exporting Harm* report of 2002. Despite the collection of locationally precise GPS data in the e-Trash Transparency Project, we do not seem to be any further along in terms of knowing whether we have a precise or accurate picture of e-waste exports. Rather than resolving disputes over the scale and scope of the e-waste problem framed as an issue of exports, the e-Trash Transparency Project has arguably heightened uncertainty rather than alleviated it. Was the earlier estimate of 50–80 percent exports wrong? Has the electronics recycling industry improved since 2002 because the GPS data, collected in 2016, indicate a range of 25–39 percent for exports? Is it changes in "market conditions" that explain the difference in findings? The e-Trash Transparency Project results implicitly raise such questions but cannot answer them.

In the academic worlds of statistical research in geography—a broad area of inquiry in which GPS is but one source of data used—there are important technical distinctions to be made between precision and accuracy. If one imagines a given phenomenon to be studied as a bull's-eye target, it is possible to employ a method that might lead to very precise but inaccurate results. To understand why, imagine a researcher is aiming to hit the target (the phenomenon to be studied) with a quiverful of methodological arrows. The arrows are released, and all of them fall in a single tightly packed cluster outside the bull's-eye. Precise but inaccurate results have been obtained. To overcome such difficulties, a researcher interested in making statistically valid claims about geographic phenomena must devise appropriate sampling strategies.

The GPS data arising from the e-Trash Transparency Project gives us precise information about particular itineraries of tagged devices and specific destinations to which they travel. Yet because of the way the study was done, we have no way of knowing whether the findings are both precise and accurate. It did not need to be this way. A differently designed study could have devised a rigorous sampling strategy that would have provided results useful to academics, activists, industry, and policy makers alike. Such sampling should account for, among other things, the variety of device types and device conditions that enter e-waste collection systems, the types of facilities (i.e., certified and noncertified) through which devices are tracked, and the geographic locations and coverage of drop-off locations in jurisdictions with and without mandated take-back legislation. Researchers involved in the e-Trash Transparency Project could have tagged devices and sent them through both certified and noncertified facilities, tracking devices to their apparent final destinations. Once data were obtained, the researchers could have reported aggregate and anonymized results. They could have gone further and alerted the individual certification governance bodies (i.e., e-Stewards and R2). Each organization could have been provided with the data covering its respective certified facilities, leaving to the certifying organizations whether to publicly disclose specific people and sites in violation of certification requirements, the law, or both. With this method, academic researchers would have a data set from which aggregated and anonymized results could be published. Certifying bodies would have precise and accurate data about the workings of their own systems, as well as the information necessary to sanction businesses found in violation of the law or their specific certification standards. And policy makers would also have access to a robust study about the character of exports that would offer better evidence for decision making. Unfortunately, such a study using GPS trackers did not come to be. The research that has been published from the e-Trash Transparency Project seems only to have magnified the uncertainty surrounding the state of e-waste exports while raising critical questions about the ethics of involving human subjects of academic research, since the e-Trash Transparency study has done much more than disclose anonymous GPS tracks.

Trade statistics, such as those I analyzed in the previous chapter, do not escape indeterminacy. All such data are limited by the accuracy of reporting to the authorities mandated to collect those data. Errors in such recording are inevitable. This is as true of trade data for e-waste as it is of data for global flows of wheat, or any other thing recorded as a commodity in trade data sets. Other inaccurate or erroneous information may enter such data sets through deliberate lack of recording or misrecording to cover instances of illegal trade (e.g., arms dealers and drug traffickers are very interested in their shipments *not* being recorded). Shipments for things categorized as waste, however, are much less black and white. It may be counterintuitive to classify wastes as something of value. In everyday experience, waste is something one typically tries to get rid of because it has come to be of zero or even negative value. Yet some businesses derive profits from managing what others want to or must discard as waste. Two of the largest waste management firms in the world are Suez Environment and Waste Management Inc. Suez achieved net sales of over U.S. $16 billion in 2015. Waste Management achieved net sales of almost U.S. $13 billion in the same year.

Other forms of waste are valued as tradable commodities. Sulfuric acid is a case in point. Though it is a waste by-product of copper smelting, it is an input into livestock feed production and thus travels across borders as a commodity. When it spills from a mine tailings pond the acid wipes out agricultural land and drinking water (see Trevizo 2014 and the next chapter for a more detailed discussion). A material generated in one industrial process might be a waste product *within* that process but be marketable as a crucial input in another industrial sector.

When we categorize, when we measure, we attempt to fix the flux of an indeterminate world. Knowing is worlding. How we world, then, matters. It would benefit discussions about e-waste to acknowledge this situation. An important step in that direction is to forgo a reactionary insistence on a truth/falsity or fact/value distinction and instead consider matters of concern. Matters of concern offer more room for analysis, inter-pretation, and negotiation in disagreements about e-waste and its geogra-phies. While those who are concerned may (and typically do) fundamentally disagree about the nature of a problem, they share concerns about it. Thinking in terms of matters of concern enables a way out of the fact/value cul-de-sac. It does so by symmetrically applying a key insight of social studies of knowledge: all knowledge is constructed, including that of the critic. Accepting that all knowledge is constructed (or built) and thus open not just to error but to fundamental indeterminacy, however, is not the same thing as claiming that all knowledge is equally fallible (see Sayer 2010, 46). When thinking in terms of matters of concern, one's staging of judgment about which knowledge claims to trust is arranged differently. Instead of adjudicating between what is taken to be (unconstructed scien-tific) fact versus (arbitrary political) value, an entirely different question is posed: since all knowledge is built, how well or badly built is this or that claim to knowledge? Asking that question means the tools, the recipes, the plans of everyone making claims about a state of affairs are put on the table for inspection by those who care to do so. Decisions about what knowledge claims to trust, then, come to be premised not on a distinction between facts and falsities or facts and values but on whether the tools used, the recipes followed, the plans consulted and instituted lead to well- or badly built knowledge claims.

If this chapter leaves readers with an antirecycling or anti-ENGO message, then it has failed. Neither of those is a missive I seek to defend. My point is entirely different. It is this: how we know e-waste matters. It matters because problem definition has concrete consequences for what and how solutions to the problem so identified are imagined and actualized. It also matters because those solutions end up affecting people and places—for good or ill. Part of what I have shown in this chapter is that e-waste is predominantly framed as a postconsumption waste management problem. This framing is done by ENGOs, but it is also a part of the industrial strategies of brand manufacturers. One of the consequences of this framing—again, a framing proffered by both ENGOs and corporate or industrial interests—is that waste arising at any point before consumers discard their devices is erased from view (see MacBride 2012, chap. 2, for a historical overview of the shift of industrial opposition to, then active promotion of, municipal recycling; see also Liboiron 2009). Yet those sites and situations prior to consumption—mining and manufacturing—are by far the largest sources of waste generated from electronics. It is to these sites and others that I turn my attention in the next chapter.

6   Weighty Geographies

Our seemingly wireless lives are predicated on a mess of tangled wires.

—Nicole Starosielski (2015, 232)

Introduction

A decade ago the popular science magazine *Discover* asked, "How much does the Internet weigh?." The answer: 0.2 millionths of an ounce, or "roughly the same as the smallest possible sand grain, one measuring just two-thousandths of an inch across" (Cass 2007). A competing analysis claimed the entire Internet is composed of two ounces of electrons (Seitz 2007; see also Krulwich 2011). Though the two figures differ by six orders of magnitude, both miss the point, according to the science journalist Robert Krulwich:

You can weigh the Internet till you are blue in the face, but the grams won't tell you anything important.

The Internet connects people. What it *is* doesn't matter. What it carries, *that matters*. Ideas aren't like chairs or tables. They have their own physics. They make their own weight. (Krulwich 2011; emphasis in the original)

Despite decades of research by humanities and social science scholars carefully documenting the materials, infrastructure, labor conditions, and energy requirements needed to make digital living possible, there is still a pernicious tendency to treat digital technologies as if they constituted a virtual, placeless and weightless elsewhere. In this chapter I push back against such tendencies that, quite wrongly, tell us that the weightiness of Internet infrastructures consists in the meaningless and neutral conduits through which the dynamism of human sociality flows. How much of that infrastructure can be removed or made nonfunctional before the ideas it carries cease to circulate partially or at all? A copper fiber telephone cable is not a fiber-optic one when it comes to transmission (Starosielski 2015). TCP/IP (the Internet transmission protocol) is not POTS (plain old telephone service). I chart some of the geographies of matter, energy, and labor needed for the creation and maintenance of digital technology and information networks so that they can run smoothly. It is a synoptic chart of digital circulatory systems and the discards or "externalities" that arise unevenly yet ubiquitously throughout those systems. The latter analogy points to a second aim of the chapter: to document the degree to which discards, remainders, and waste from digital technologies arise unevenly but at all points of their existence, not merely when they are thrown away. Different infrastructures afford different modes of living together and different modes of discarding.

There are two main reasons why I document the scale and location of waste arising from digital technologies ubiquitously but unevenly throughout their existence. First, it is crucial to demonstrate that resource extraction for, and the manufacturing of, electronics generate vastly more waste than does the postconsumption discarding of gadgets. Second, I want to more fully develop a point raised in earlier chapters, namely, that the dominant story line about e-waste being a postconsumer waste management problem is a partial, even peculiar, framing of the problem of e-waste since it typically leaves out the vastly higher amounts of waste arising from resource extraction and the manufacturing processes necessary for the creation and use of digital devices. No amount of postconsumer recycling will recoup discards arising upstream in resource extraction and manufacturing before the devices make it into retail consumers' hands. The importance of the latter point is further developed by noting how bracketing out waste that arises during resource extraction, manufacturing, and use guarantees that the e-waste problem cannot be solved through postconsumer recycling. Yet just such recycling, derived from the dominant story line about e-waste, is the prevailing policy prescription for managing it.

Since the vast proportion of waste is generated before consumers even purchase their devices, let alone use them, the hope that recycling electronics will reduce e-waste is a false one. In addressing this topic, this chapter deepens the analytical trope of defamiliarization of electronic discards that is important to my overall argument. It also plays on the notion of "weightiness" in the double sense of the massive energy and material consequences of waste arising in resource extraction, manufacturing, use, and disposal of electronics and of the importance of those consequences in their highly uneven distribution among people and places.

Discardscapes and Harm

Geographers distinguish between space as a thing or process and spatiality, which denotes the qualities of space. Spatiality captures the idea that spaces are brought into being, rather than merely existing ready-made, and in turn have their own generative effects. Spatiality also denotes the embodied and aesthetic characteristics of space. Tuning these meanings to the study of landscapes has in many ways been core to geographers' approaches to Earth as an object of analysis.

The combining form -scape denotes a certain kind of scene. In this chapter I turn my attention to a variety of such scenes as they relate to the existence of electronics. These are scenes of resource extraction, of manufacturing, and of use, or what I call "minescapes," "production-scapes," and "clickscapes." In turn, each of these scene types assembles discardscapes. The concept of discardscape represents an attempt to capture three interconnected ideas. The first idea is that the various scenescapes comprise sites and situations assembled into geographies with particular spatialities. One of the principal qualities of their geographies is that they tend to be omitted from stories about e-waste. The second idea is that discards from electronics arise ubiquitously but unevenly at all points of their existence, not just when they are thrown away. This second point leads to a third and corollary point about discardscapes more generally, not just those that pertain to electronics. It is this: because of the fundamental indeterminacies about waste from which we cannot escape, it might be useful to think of discardscapes as patchy, distributed and not necessarily coherent geographies. But noncoherence is different from incoherence. The former is, in principle at least, partially knowable, whereas the latter is not (Law 2004b). Just because discardscapes do not necessarily cohere into a knowable totality, it does not follow that action for change is futile or that it should be indefinitely deferred while we await certainty and business as usual continues apace.

Understood with these three points in mind, then, discardscapes are a type of scene in which storytelling is about recovering the geographies bracketed out by prevailing narratives about waste as an externality to, rather than a fundamentally constitutive element of, industrial ways of life (Liboiron 2013a). I chart the discardscapes of electronics to elaborate the central argument of this chapter: in terms of weight, discards arising from mining, manufacturing, and use vastly exceed those that occur when consumers throw away digital devices. Framing e-waste as a postconsumption problem guarantees that proposed solutions will fail to reduce the vastly more significant wastes that arise from the production and use of these technologies. The geographies charted in this chapter, then, are weighty in multiple senses of the term—in their mass, but also in their consequential importance.

Samantha MacBride (2012) argues that contemporary waste can be distinguished from its historical antecedents by the tonnage, toxicity, and heterogeneity of the materials that compose it. Electronic waste certainly meets MacBride's criteria for being a contemporary form of waste. Charting its weighty geographies necessitates adding concepts of harm that exceed toxicity (Liboiron 2015). When it comes to the potential for harm over very long, effectively permanent, time horizons we tend to think of nuclear waste as special in this regard (indeed, nuclear waste is often regulated as a "special waste"). Yet it is not. Many segments of contemporary waste streams are effectively permanent (Gray-Cosgrove, Liboiron, and Lepawsky 2015), as are the effects of their common mode of disposal in landfill (Clark and Hird 2013). As I chart the weighty geographies of electronic discards in what follows, the horizons of space and time usually associated with nuclear waste will become part of the picture. Certain gases used and released in the manufacture of electronics not only have global warming potentials thousands of times higher than that of CO_2 but also have atmospheric residence times ranging from 270 to 50,000 years. The plastic materials of which electronic devices are composed break down on geological time scales and indeed are becoming incorporated into new forms of geological strata (Corcoran, Moore, and Jazvac 2014).

Minescapes

Scenes of mineral extraction rarely enter e-waste discussions except in very broad terms about resource consumption (Smith, Sonnenfeld, and

Pellow 2006). But at the turn of the twenty-first century a sudden jump in price for tantalum turned public attention to the mining frontier in several African countries, especially the Democratic Republic of Congo (DRC). Common explanations for the price jump pointed to demand from electronics manufacturers, especially due to booming demand for cell phones and other digital devices. Activism and concerned photojournalism about the mining of tantalum, tin, tungsten, and gold (known as 3TG, and also as "conflict minerals") in Africa took off. Attention-grabbing images of child laborers in mines extracting minerals showed readers of *National Geographic*, for example, that "our electronic devices have bankrolled unspeakable violence in the Congo" (Gettleman and Bleasdale 2013).

The timing of the price jump, beginning as it did in late 2000, corresponded with what would in retrospect be known as the Internet or dot-com bubble. The early 2000s also saw the emergence of concerns about a looming supply crisis for 3TG minerals, which resulted in growing speculation and rising prices. But as the geographers Christoph Vogel and Timothy Raeymaekers (2016) show, the story is far more complex. Demand from the electronics sector for 3TG minerals actually plays a relatively small role in aggregate demand for these minerals. Up until 2000 the tantalum market was dominated by a small cartel of processing firms. The rise in prices both threatened the stability of this cartel market for strategically important minerals and opened opportunities for new players to profit from this instability. Into this instability stepped the U.S. Defense Logistics Agency (DLA), which had substantial stocks of tantalum in its possession. In December 2000 the DLA sold 250,000 pounds of tantalum ore at a price some ten times higher than the prevailing prices of only a few months earlier. This ore was bought by Kamco Metals, a hitherto unknown player in the tantalum trade. Kamco, it seems, may have been acting as a front for business interests in Kazakhstan, where a major refining facility, the Ulba Metallurgical Plant, is located. The players behind Kamco may have seen market instability as an opportunity to profit from the shakeup of the prevailing tantalum cartel. The DLA may have been acting to quell world market prices for this strategic metal as well as to restructure the international market toward U.S. strategic interests. One of the effects of the DLA sale to Kamco was that an important member of the tantalum cartel, Belgium's Umicore, lost market share to U.S.- and Kazakhstan-based competitors (Vogel and Raeymaekers 2016).

As with the regulation of hazardous waste, indeterminacy is relevant to understanding the relationships between the minescapes of 3TG minerals and the generation of discards from electronics. Whatever the motivations were for the maneuvers in the tantalum markets by the DLA, Kamco, and others, public concern over links between 3TG minerals and consumer electronics grew into major NGO campaigns, including "No Blood on My Mobile" and calls for trade embargoes against 3TG minerals sourced from the DRC (Vogel and Raeymaekers 2016). These concerns were expressed strongly enough to be translated into U.S. legislation. Section 1502 of the U.S. Dodd-Frank Act requires firms to disclose the use of conflict minerals in their supply chains that originate in the DRC (U.S. Congress 2010, 839). Vogel and Raeymaekers (2016) show that the Dodd-Frank Act is concerned with clamping down on extortion by armed nonstate groups but not with controlling extortion by national armies in the DRC or adjacent countries. The effects of the combination of market restructuring and embargoes written into the Dodd-Frank Act ended up devastating the artisanal mining sector in the DRC. Some 500,000 such miners and their six to eight million dependents have seen their livelihoods vanish as 3TG buyers turned elsewhere (Vogel and Raeymaekers 2016). Efforts to reform the 3TG supply chain are thus far from being an unmitigated success since they have increased the vulnerability of the very people these reforms are supposed to have protected and ignore the role of demand for these minerals in China and other emerging markets (Nest 2011).

A recent study collected all filings with the Securities and Exchange Commission (SEC) of conflict mineral disclosure reports mandated by the Dodd-Frank Act (Y. H. Kim and Davis 2016). Of the more than 1,200 individual companies that submitted the required forms in 2015, "eight out of ten admitted they were unable to determine where the raw minerals contained in their products originated … [and] only 1% of companies declared their products to be free of any conflict minerals with great certainty" (Y. H. Kim and Davis 2016, 1906). The primary reason for these results is the dispersed character of supply chains in the different industry groups captured in SEC filings. The presence of such dispersed supply chains is especially relevant to the electronics sector. Electronics brand manufacturers typically rely on multiple tiers of contract suppliers, manufacturers, and assemblers, making it very difficult to have certainty about supplies of materials. Even the socially minded initiative Fairphone, which explicitly attempts to design and build "fair trade" phones free of conflict minerals, notes uncertainty about the origins of some minerals used in its products despite its own field investigations of mines in East Africa that are part of the Fairphone supply chain (Fairphone 2013; Winter 2013).

Supply chain complexity, along with differing definitions of both conflict minerals and regions of conflict, ramifies the indeterminacy of attributing specificities of particular minescapes to the electronics industry. One recent study estimates the electronics sector consumes up to 15 percent of the global supply of tantalum and up to 5 percent of all other 3TG minerals (Fitzpatrick et al. 2015, 979). So even while the electronics sector has become the focus of several activist campaigns against the use of conflict minerals (e.g., Global Witness 2017; Partnership Africa Canada 2017; RAISE Hope for Congo 2017; see also Gettleman and Bleasdale 2013), some 85–95 percent of industrial consumption of 3TG minerals occurs in other sectors. The point is not to imply that the electronics sector is an insufficiently significant consumer of these minerals to bear some responsibility for the human and environmental discards of 3TG minescapes. However, the inherent indeterminacy in deciding at what threshold the electronics sector's consumption becomes too much is inescapable. Some may argue that any amount is too much. Others may make a case for a greater amount. But there is nothing in the physical properties of 3TG minerals that provides the fixed moral bedrock on which to base such a determination. Instead, as Fitzpatrick and co-workers write (2015, 979), "Simple narratives around a single sector create a risk for not fully addressing the problem" (see also Jameson, Song, and Pecht 2016; Nest 2011). It may actually be possible for consumer electronics to be made without 3TG minerals. But if that comes to pass, a very large percentage of the industrial consumption of those minerals will remain unchanged.

* * *

I am standing on the lip of the Lavender Pit, a gaping hole in the Earth some 1.2 km^2 in area and 274 meters deep. At the bottom is a small lake colored lurid shades of vermillion, sienna, and orange (see figure 6.1). The Lavender Pit is a former mine that is part of the larger Copper Queen Mine complex. Just a few minutes' drive east of Bisbee, Arizona, on AZ-80 is a pull-off where motorists can step out of their car and onto a small outcrop for a view of the pit from its northern rim. Here there is also a plaque that

Figure 6.1
View of the Lavender Pit, which is part of the Copper Queen Mine complex near
Bisbee, Arizona. Between 1975 and 2003 the U.S. electrical and electronic manufac-
turing sector annually consumed about the same amount of copper as was extracted
from the Lavender Pit over its entire twenty-four-year operational life.
Source: Author's photograph.

charts the rise and fall of demand for copper over time. In an ironic twist,
1975, the year after the Lavender Pit ceased operation, is marked as a point
of rising demand for copper, coinciding with the "first personal computers"
(see figures 6.2 and 6.3).

Over its operational life as a mine, from 1950 to 1974, the Lavender Pit
produced about 544,000 metric tons (approximately 600,000 U.S. tons) of
copper. For comparison, the domestic U.S. electrical and electronic products sector consumed, on average, about 593,448 metric tons of copper per
year between 1975 and 2003 (U.S. Geological Survey, Kelly, and Matos
2005). Roughly the amount of copper extracted over the twenty-four-year
production life of the Lavender Pit was being consumed by electronics
manufacturers every year in the United States alone. Producing those
544,000 metric tons of copper from the Lavender Pit entailed "more than
256 million [U.S.] tons of waste" being stripped from the mine site (Long
1995, 38; U.S. Geological Survey 1991). That works out to an annual

Figure 6.2

Information plaque at the roadside lookout over the Lavender Pit. It notes in part that "every electronic gizmo from refrigerators to iPods needs copper wiring."
Source: Author's photograph.

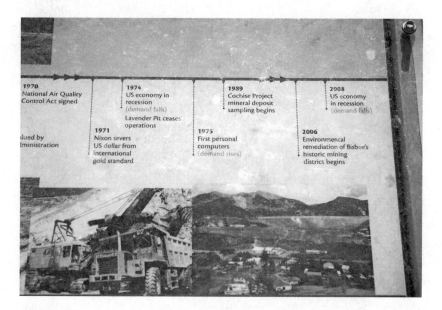

Figure 6.3

A close-up view of the information plaque at the Lavender Pit. The year 1975 marked a rise in demand because of the growth in personal computing. The Pit had closed the year before, in 1974.
Source: Author's photograph.

average of over 10.6 million U.S. tons of waste arising from the Lavender Pit. In other words, for every one ton of metal extracted, 426 tons of waste arose. That ratio may seem extraordinary, but it is not at all unusual in the mining industry, where a typical operation "yields greater than 98 percent waste from the excavated material" (Keeling 2012, 554).

Though the U.S. Geological Survey refers explicitly to "waste" from the Lavender Pit, that terminology is more nebulous than it might seem—what were mine wastes at one time can become valuable again with changing economic or technological circumstances (Lottermoser 2003). After the Lavender Pit ceased being a working mine, its wastes were reanimated as valuable materials through the use of hydrometallurgical techniques and, later, patented bioleaching technology. The plaque at the highway pull-off indicates mining ended at the Lavender Pit in 1974, but the mine complex did not actually stop producing copper and other base metals that year. Leaching processes were applied to waste dumped from the mine and to the walls of the open pit to recover small amounts of copper and other metals right up until 2002. After that a joint venture between a biotechnology company and the mine's owner led to the construction of a bioleaching plant at the site, and the patented process enabled copper extraction from dumped mine waste near the site until 2013 (Briggs 2015).

This resuscitation of former mine waste did not mean its historical discards went away or that new ones were not generated. Material from the Lavender Pit was dumped on the outskirts of Bisbee and nearby Warren (see figure 6.4 online). The U.S. EPA's Toxic Release Inventory contains data for releases from Copper Queen between 1998 and 2005. Total on- and off-site releases amounted to over 1,700 metric tons of chemicals over those seven years. Zinc and copper compounds accounted for the vast majority of these releases, at approximately 625 metric tons and 1,000 metric tons, respectively. The remainder consisted of sulfuric acid and a small amount of lead.

The discussion of the Lavender Pit draws attention to the discards of just a single mining complex related to copper. Copper is a widely used industrial input, so only a small amount of the total copper produced from this complex eventually entered the manufacture of electronics. Yet the scale of the discards from the Copper Queen mines is instructive for thinking about postconsumer e-waste as a matter of concern. The annual average waste generated from the Lavender Pit mine before it closed exceeded what the UN StEP program estimates was the total amount of e-waste arising in the entire United States in 2015 (United Nations StEP 2017).

About 70 kilometers south of the Lavender Pit is a site where I conducted fieldwork. It is a small town in northern Mexico where a small electronics repair and recycling business was trying to sustain itself in a region from which manufacturing firms had departed, gutting local employment options. A copper smelter operated by Grupo México is located less than an hour's drive from this recycling business near the town of Nacozari, Mexico. For years, this repair and recycling operation attempted to form a working relationship with that smelter by supplying nonrepairable CRTs as an input to the copper smelting process (Lepawsky et al. 2017). CRTs contain lead and silica, which are useful as fluxing agents in that process. The smelter has a reported capacity of 1 million metric tons per year (Grupo México and Southern Copper 2014). The electronics repair and recycling operation was proposing to sell less than 500 metric tons of CRTs per year to the smelter. Among other benefits, the sale of nonrepairable CRTs to the smelter would have meant they could be recycled without being shipped to the United States for further processing (thus also reducing emissions from transport) and would also have reduced the smelter's reliance on mined lead and silicate, thereby cutting down on the negative environmental effects of mining operations while also supporting local employment in the town. However, CRTs occupy a regulatory gray zone. Under some interpretations of U.S. and Mexican laws, CRTs count as hazardous waste and thus require special permitting on the part of trade partners. The repair and recycling operation eventually won permission to move the CRTs to the smelter, but the smelter ultimately refused to enter the partnership because doing so would have required that the facility become a permitted hazardous waste handling facility, thus increasing its operating costs and regulatory oversight of it. A study of this same smelter at Nacozari by the U.S. Geological Survey calculated that in 2004, it generated discards comprising 650,000 metric tons of slag, 986,000 metric tons of sulfuric acid, 29,000 metric tons of sulfur dioxide, and 64,000 metric tons of CO_2 (U.S. Geological Survey and Goonan 2005, 81). The most recent data available from the smelter operator provides figures only for sulfuric acid production. These discards amounted to 819,000 metric tons in 2011, 887,800 metric tons in 2012, and 719,500 metric tons in 2013 (the latest year for which data are available, see Grupo México and

Southern Copper 2014). In other words, in any one of the years for which data are available, the discards of sulfuric acid alone from this one smelter were equal to between 75 and 103 percent of what the UN's Solve the E-waste Problem (StEP) initiative estimates was the total amount of postconsumer e-waste arising in the entire country of Mexico in 2014 (United Nations StEP 2017a). Total discards from the one smelter in 2004 calculated from the USGS report amount to about 1.7 million metric tons, just shy of one quarter of the seven million metric tons of postconsumer e-waste arising in all of the United States in 2014 (United Nations StEP 2017b).

Unlike the CRTs, the sulfuric acid generated at the Nacozari smelter is not regulated as hazardous waste—even though it is a by-product of smelting copper that the smelter does not want—because under the relevant regulatory frameworks, sulfuric acid is classified as a commodity (it can be used in the manufacture of livestock feed, for example; see Möller 2009). That waste sulfuric acid arising at the smelter is transported by train and truck at grade right through the center of the small town where the electronics repair and recycling operation is located and where I was doing fieldwork. Trucks with warning glyphs for acid came through the town daily while I was there. Late at night, roughly 50 meters from where I slept, I could hear the rumble of the trains carrying the acid north across the border into the United States.

Between August 6 and 7, 2014, some 41.6 million liters (approximately 11 million gallons) of copper sulfate acid spilled from Grupo México's Buenavista del Cobre open-pit copper mine, about 60 kilometers east of the town where I was doing fieldwork. The Buenavista mine is one of the mines feeding the smelter at Nacozari. The spill entered the Bacanuchi river and from there fed into the larger Sonora river (Trevizo 2014). As a consequence of the spill, 25,000 people lost access to safe water, crops were spoiled, and rangeland lost the ability to support cattle, one of the region's major sources of income.

The density of a given solution partly determines its mass. If the 41.6 million liters spilled from the mine were pure water, the accident would have resulted in a release of 41.6 million kilograms (or approximately 101,633,102 pounds) of liquid into the river system. By that approximation, this one spill from a single mine site that occurred over a twenty-four-hour period released about 0.4 percent of the mass of postconsumer e-waste calculated to arise annually in the entire country of Mexico (the UN's StEP [2017a] initiative calculates 958,000 metric tons of e-waste arose in Mexico in 2014).

Productionscapes

Manufacturing electronics involves materials and energy-intensive processes, from which substantial discards arise. For example, making a desktop PC generates CO_2 emissions equivalent to those arising from almost two years of using that computer, assuming it is in use five hours a day, seven days a week (Puca et al. 2017; my calculations from figures in table 5, p. 133). To produce a single 2 gram microchip requires at least 72 grams of chemicals, 1,600 grams of fossil fuels, and 41 megajoules of energy (Williams, Ayres, and Heller 2002). This makes the intensity of resource use for chip fabrication two orders of magnitude larger than that for cars.

Back in 2004, when CRT monitors were still standard features of desktop computers, it took approximately 6,400 megajoules of energy and 260 kilograms of fossil fuels to make the computer-monitor pair (Williams 2004, 6166). Around 2009, shipments of flat panel displays (FPDs) surpassed those of CRTs (DisplaySearch 2014). Monitors for computers got noticeably thinner and lighter, but discards arising from their production have hardly gone away. The manufacture of FPDs requires the use of a class of fluorinated greenhouse gases (F-GHGs), all of which have global warming potentials thousands of times higher than that of CO_2 (U.S. Environmental Protection Agency 2016a). F-GHGs are used in cleaning and pattern etching applications in the FPD manufacturing process. The atmospheric lifetimes of these gases range from 270 to 50,000 years, depending on which specific chemical is considered. High degrees of uncertainty exist over the relative contribution of the FPD sector to F-GHG emissions as preliminary studies of it have only just begun. However, aggregate emissions from the sector amount to 1.75 million metric tons carbon equivalent (MMTCE) (U.S. Environmental Protection Agency 2016b, 4; Ohkura et al. 2012). Data that do exist indicate that 78 percent of those total emissions consist of sulfur hexafluoride (SF_6) (Electronic Industry Citizenship Coalition 2016, 3). SF_6 has an atmospheric lifetime of 3,200 years and a global warming potential 22,800 times higher than that of CO_2 (U.S. Environmental Protection Agency 2016a). Some firms in the FPD market are switching to nitrogen trifluoride (NF$_3$), which has a lower global warming potential than SF$_6$. However, with a global warming potential 17,200 times that of CO$_2$ and an atmospheric residence time of 740 years, NF$_3$, though representing an improvement over the use of SF$_6$, still has very substantial effects (U.S. Environmental Protection Agency 2016b).

Waste arising from electronics manufacturing vastly exceeds that arising from postconsumer discards. Data made available by some manufacturers show this claim to be uncontroversial. For example, Apple publishes data on the carbon footprint of its products. Its data show the first version of the iPhone entailed a total of 55 kilograms of CO$_2$ equivalent (CO$_2$e) released per device. Total emissions associated with the phone rose over the next several models, peaking with the iPhone 6 at 110 CO$_2$e (a 100 percent increase in total emissions compared to the first iPhone model). Since the iPhone 6 total emissions have declined again with the iPhone 7 and 7 Plus, at 56 and 67 kilograms of CO$_2$e, respectively. Total emissions, however, are only part of the story. It is crucial to consider what points in the production and postproduction lifetime of these devices proportionally account for total emissions released. Apple's data are categorized into production, transport (i.e., getting the product to market), use by the consumer, and recycling (or end of life). Looking at the data by category discloses that production is by far the largest contributor to CO$_2$e emissions. The only device in which production accounted for less than half of total emissions was the first model of iPhone (see figure 6.5). In every model since then, production has accounted for a growing share of total emissions.

What does this mean? First, it does not necessarily mean that Apple's manufacturing processes are getting less efficient (and releasing more CO$_2$e). Instead, the increasing proportion of CO$_2$e associated with production likely results, at least in part, from efficiency gains in the use phase (e.g., as the phone's power management becomes more efficient). But there is another implication: recycling the iPhone at its end of life will not, by definition, recoup any of the emissions that were released *before* a consumer bought the phone and started using it (i.e., emissions associated with production and transportation). Indeed, as Apple's own data suggest, recycling *adds*, however trivially, to the emissions associated with the phones.

My point is not to vilify Apple (I wrote this book on an Apple device). Apple and all the companies that publish data on the environmental foot-

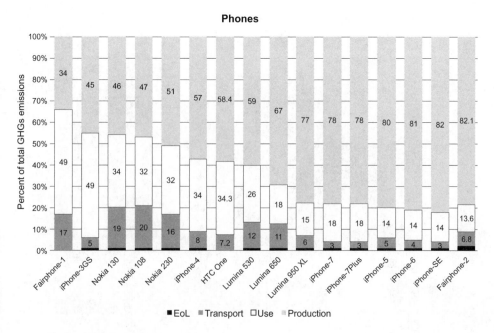

Figure 6.5
Percent of total greenhouse gas emissions released in the production, transport, use, and at end of life of various models of phones.
Source: Data were sourced from company reports. Calculations by the author.

print of their devices should be applauded. And those that do, show patterns very similar to those evident in the data for iPhones. Production as the point where the largest share of emissions occurs is not a characteristic only of Apple devices. To be sure, among manufacturers there is variability in the proportion of CO_2e attributable to production. But, generally speaking, the newer the model, the more that production contributes to total emissions. This is the case for phones as well as for other major classes of consumer electronics such as laptops (see figure 6.6), desktops (see figure 6.7), and tablets (see figure 6.8). Postconsumer recycling cannot cancel out emissions released in production, transport, and use.

* * *

The tonnage of discards arising from the production of electronic devises is important, but so are their chemical heterogeneity and their toxicity. The tonnage, toxicity, and heterogeneity of discards from the production

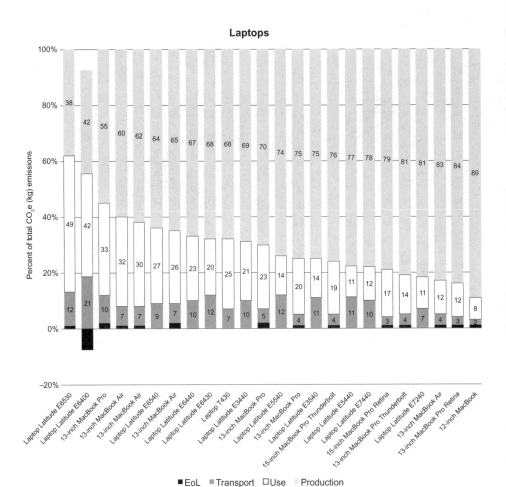

Figure 6.6
CO_2e released in the production, transport, use, and at end of life of laptops.
Note: Dell assumes its E6400 model laptop is made of 75 percent recycled materials and therefore assumes recycling gives a 30 kilogram CO_2e credit, to total CO_2e released.
Source: Data were sourced from company reports. Calculations by the author.

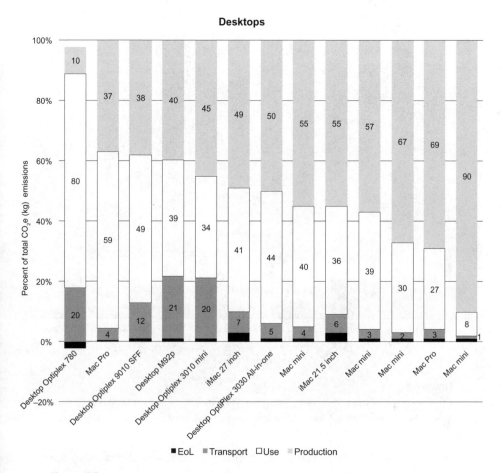

Figure 6.7
CO_2e released in the production, transport, use, and at end of life of desktops. *Note*:
Dell assumes that the Optiplex 780 model desktop is made from 75 percent recycled
materials and thus assumes that recycling provides a 20 kilogram CO_2e credit, to
total CO_2e released.
Source: Data were sourced from company reports. Calculations by the author.

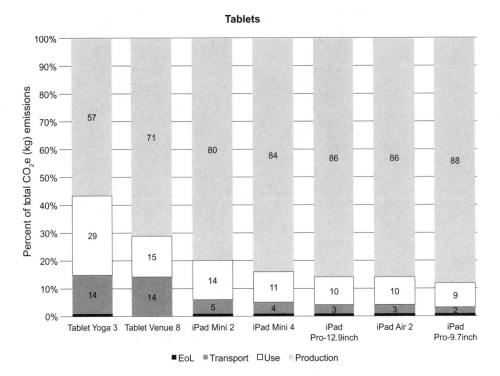

Figure 6.8
CO$_2$e released in the production, transport, use, and at end of life of tablets.
Source: Data were sourced from company reports. Calculations by the author.

of electronic devices thread through multiple sites and diffuse through multiple spheres of Earth. Extant data documenting these discardscapes are far from complete, but those that do exist start to give some sense of the location, scale, and scope of their effects. Silicon Valley is a good place to look.

Evidence of toxicological harm to workers in semiconductor industries began to be published in the early to mid-1980s (LaDou 1984, 1986; Pastides et al. 1988). These harms were disproportionately incurred by the bodies of production and maintenance employees, particularly women. Together these staff made up approximately 31 percent of total employees in the industry, yet they experienced 60 percent of total injury and illness cases reported in the sector. Occupational illness among semiconductor workers in California was found to be three times higher than in the general manufacturing industry (LaDou 1991). Broad indications of harm to reproductive health and of "excessive risks" for some forms of cancer and risks to women's reproductive health are also found among workers in this sector (I. Kim, Kim, and Lim 2015; M.-H. Kim, Kim, and Paek 2014, 95). Still, indeterminacy haunts even quite recent occupational health research on harm to electronics workers. The broad findings are muddied by substantial limitations in epidemiological research. These limits currently make it difficult to confirm or rule out causal links between all forms of chemical exposure of workers in electronics manufacturing and diseases such as cancer (I. Kim et al. 2012). While the chemical sources of these harms do not make it into the final products of electronics manufacturing and enter bodies where they are not intended to be, they rarely, if ever, fall under the problem of "e-waste" as it is predominantly narrated. That this is so tells us at least as much about the prevailing narratives about e-waste and those who narrate them as it does about e-waste itself.

Harm entered workers' bodies at their place of employment, but a broader population was also at risk from chemical discards from the industry through groundwater contamination (Olivieri et al. 1985). In the 1970s the Santa Clara Center for Occupational Health (SCCOSH) formed an advocacy project called the Silicon Valley Toxics Coalition (SVTC, which later became a separate organization) to articulate these concerns and seek ameliorative action (Smith, Cole, and Wilmsen 2003). SCCOSH action led to investigations by the U.S. EPA and, eventually, the designation of more than twenty Superfund sites in the valley (Gabrys 2011; Smith, Sonnenfeld, and Pellow 2006; U.S. Environmental Protection Agency 1989, 3). Many of these Superfund sites remain active zones of cleanup today (see figure 6.9 online).

The geographies of electronics production networks have been rearranged substantially over time, but the legacy of discards from their manufacture in Silicon Valley remain active, along with production facilities still in operation there. A principal concern is groundwater contamination resulting from historical releases of chemicals used in electronics manufacturing processes at firms such as Fairchild Semiconductors, Raytheon, and Intel in an area known as the Middlefield-Ellis-Whisman (MEW) Study Area (U.S. Environmental Protection Agency 2017a). Contaminants of concern include trichloroethane (TCE) and the chemicals it degrades into over time. After extensive study, the U.S. EPA characterized TCE as "carcinogenic in humans by all routes of exposure" (U.S. Environmental Protection Agency

2011, xlii). TCE also poses noncarcinogenic hazards. These include toxic effects on the central nervous system, the immune system, kidney, liver, the male reproductive system (with more limited evidence for the female reproductive system), and the gestating fetus.

Mitigating the effects of TCE and other contaminants associated with Superfund sites in Silicon Valley employs pump-and-treat (P&T) systems. These are a general class of remediation technologies used to filter water contaminated with dissolved chemicals. In essence, P&T systems siphon groundwater through extraction wells sunk into areas of contamination (U.S. Environmental Protection Agency 2012c). The extraction wells pump water to the surface, where the water moves into treatment infrastructure. The basic treatment process uses activated carbon filters or air stripping to remove contaminants. Activated carbon filtering entails the use of carbon-rich material acting as a filter to which contaminates stick (U.S. Environmental Protection Agency 2012a). In contrast, air stripping forces air through contaminated water in an aeration tank and the air evaporates the contaminates into collection and treatment infrastructure (U.S. Environmental Protection Agency 2012b).

While P&T is simple in principle, it entails a range of complexities that condition its effectiveness. The type, concentration, and mixture of chemical contaminants, the underlying geology and hydrology, and the size of the contaminant plume are among the factors that influence the effectiveness of P&T systems. Depending on the mix of factors unique to a given site at which P&T is being used, the effectiveness of this technology can decline over time. A presentation by environmental engineers to the Federal Remediation Technologies Roundtable, a consortium of U.S. government agencies, indicated that a business-as-usual scenario of P&T remediation at MEW would require 700 years to reach target levels (Gallinatti et al. 2012, 27; see also Rust and Drange 2014). Moreover, the P&T process does not eliminate the contaminates it is used to clean up. It also generates its own discards.

The P&T systems in use in Silicon Valley do not themselves eliminate the contaminants they collect. Instead, they move those contaminants out of the effected groundwater systems and concentrate them in the filtering apparatus of the P&T system. This means that the filters themselves become saturated with the contaminants, which mandates regular filter changes. Used filters then require their own treatment as hazardous waste at specialized facilities, and there are only a few such facilities around the country licensed to handle them. As investigative journalists Susanne Rust and Matt Drange (2014) found, the journey from Silicon Valley to such hazardous waste facilities often means thousands of kilometers of travel by truck to facilities on the other side of the country. In effect, the situation means P&T cleanup activity itself also generates discards. Rust and Drange (2014) calculate that for every 5 pounds of contaminant extracted from Silicon Valley sites for treatment and off-site disposal, about 20,000 pounds of CO_2 arise. Waste arises from the treatment of waste.

Electronics manufacturing continues in Silicon Valley, indeed nationally in the United States, though much of the sector has moved offshore over time (Lüthje 2007). Toxic Release Inventory (TRI) data show that more than 194 million pounds (more than 97,000 U.S. tons) of chemicals deemed hazardous by the U.S. EPA were released by the computer and electronics sector nationally between 1991 and 2015 (U.S. Environmental Protection Agency 2017c). The three most important releases by weight are freon 113 (more than 26.5 million pounds), an ozone-depleting gas that can also cause nervous system damage; 1,1,1-trichloroethane (more than 26.4 million pounds), another ozone-depleting gas that can cause liver and nervous system damage; and methyl ethyl ketone (more than 20 million pounds), which can cause harm in utero (see figure 6.10).

Almost 17 million pounds of total releases occurred in California, followed by Alabama, with more than 15.5 million pounds, and Oregon, with more than 14.7 million pounds (see figure 6.11). Indeed, mapping the different types of chemicals released at the Zip Code level across the United States offers a cartography of weighty geographies we forget when waste from electronics is axiomatically understood as that which is discarded after use (see figure 6.12 online).

Overall, hazardous chemical releases from the computer and electronics sector in the United States have declined substantially over time as the industry has moved offshore (see figure 6.13). But offshoring does not eliminate those releases. It moves them elsewhere, to the geographies of contemporary ICT production networks, mostly in Asia and particularly in China. The chemical harms experienced historically by workers in Silicon Valley have been largely exported with the overseas supply chains for contemporary electronics production (Bloomberg 2017; Simpson 2017).

Figure 6.10
Chemical releases from the U.S. computer and electronics sector.

Chemical releases in U.S. and Puerto Rico (total pounds)

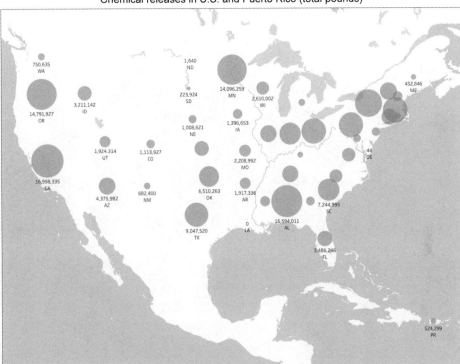

Figure 6.11
Total chemical releases from the U.S. computer and electronics sector by state.

It is much more difficult to account for the ubiquitous but unevenly distributed discardscapes of electronics globally since no data equivalent to those of the EPA's TRI database exist (though see U.S. Environmental Protection Agency 2017b for some ongoing international initiatives to this end). Even the TRI data that do exist have important limitations. For example, they record only hazardous waste releases by industry, but not measures of nonhazardous waste arising. As a consequence, TRI data vastly understate total amounts of waste arising from the electronics industry. By how much is impossible to know under current data collection regulations (MacBride 2012, 90–91). There is also evidence of inconsistent reporting of data within the TRI database itself at least for the industrial sector analyzed here. Figure 6.14 (online) illustrates that some reporting of some chemical releases from the sector is discontinuous in time. Whether these

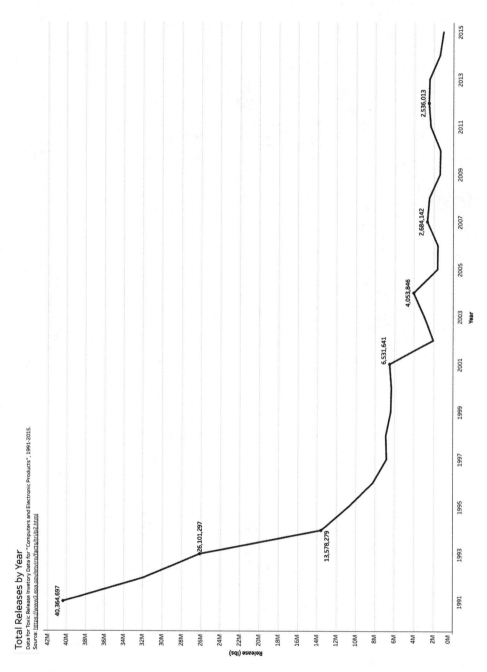

Figure 6.13

Total chemical releases from the U.S. computer and electronics sector, 1991–2015.

discontinuities represent inconsistent reporting, changes in industrial practices that phased out some chemicals, or some other factors is not clear
from the data themselves. Under such conditions of indeterminacy, it is
only possible to guess at an accurate portrayal of the weighty geographies
of discarded electronics when supply chains for the industry are distributed
globally but highly unevenly (Grimes and Sun 2016).

Clickscapes

The notion of the Internet as a singular, well-ordered whole—topologically
networked, geographically agnostic, and weightless—is a story that depends
on three particular ways of staging it. The first presumes a singular wholeness that is actually absent from the situation at hand. Strictly speaking,
there is no such thing as *the* Internet. Instead there are indeterminate
arrays of multiple networks. Yet "we typically internalize the fiction that
there is only one Internet, and one cloud" (Hu 2015, xxi). A second corollary follows from this absence of wholeness from the scene: we are unable
to know within any reasonable bounds of certainty how much and what
types of energy are consumed by Internet infrastructure. Such "known
unknowns" remain difficult to pin down, for a variety of reasons (Horner,
Shehabi, and Azevedo 2016). Among the known unknowns are a lack of
data on how people use different network enabled devices and wide varia-
tions in the definitions of what sorts of infrastructure and what use conditions should be incorporated into such analyses. A third staging also lends
coherence to the story. This is the bracketing out of the multiple discardscapes that arise from the use of networked digital devices. If those discardscapes are charted, however, stories about the Internet as a weightless,
well-ordered whole do not stand up to scrutiny.

Data centers account for an appreciable and growing portion of worldwide energy use, estimated at between 1.1 and 1.5 percent of total electricity consumption (Koomey 2011; Koomey, Matthews, and Williams 2013).
Worldwide growth in energy use by data centers has expanded rapidly.
Between 2000 and 2005, such energy use expanded at an average annual
rate of more than 16 percent, though growth slowed between 2005 and
2010 to just over 11 percent annually. The fastest rates of growth, between
2000 and 2005, were attributable to the expanding economies of China,
India, and other Asian economies (excluding Japan). Data centers consume

electricity both directly in the operation of the servers they house and in the infrastructure necessary to keep those servers cooled and supplied with backup power. It is roughly evenly split between these two types of electricity consumption keeping the servers cold and provided with backup electricity (Koomey 2008, table 2, p. 4; 2011, figure 1, p. 13). Of course, the many different ways in which this electricity is generated condition the discardscapes associated with energy use in data centers. Coal, natural gas, nuclear, hydroelectric, wind, and solar power each generate their own discardscapes. Yet just because digital technology companies are making significant (and laudable) investments in renewable energy sources (Greenpeace 2017) does not mean problematic discards disappear. For example, reservoir flooding for hydroelectric development can mobilize the potent neurotoxin methylmercury into food webs—a process that has a direct and disproportionate impact on indigenous people in the Canadian province I have settled in (Schartup et al. 2015). When the project is complete, it "will power homes and businesses across Newfoundland and Labrador with clean, renewable energy for generations to come" (Nalcor Energy 2017), but at a neurotoxic cost to the people most dependent on the land for food.

Common practice is to think through the environmental impact of ICT use in terms of carbon emissions, notwithstanding the substantial challenges of indeterminate system definitions and boundaries. Malmodin and co-workers (2010) estimate a worldwide sectoral total of 969 MMTCE for the operation (i.e., use) of ICTs and electronic entertainment devices (see table 6.1). A relatively small but appreciable portion, roughly 2 percent, of total global emissions of CO_2e can thus be attributed to the clicks we collectively make when using our digital devices. This figure is much more illustrative than precise. But precision is not the issue here. Instead, the point is that discards arise at all points in the production and lifetime of electronics, not just when they are thrown away by consumers. Moreover, it remains unclear whether there is any good reason to think that the growing use of ICTs will offset the energy and materials consumed to make and distribute products and services that ICTs replace (Galvin 2015; Horner, Shehabi, and Azevedo 2016).

More than 95 percent of telecommunications travel through undersea cables (United Nations Environmental Programme 2009, 3). About 428 such cables were in service in early 2017, amounting to over 1 million

Table 6.1

Energy use by selected ICTs.

ICT	Operational electricity (TWh)	Total CO_2-eq emissions (Mt)
Mobile networks, operation	50	46
Mobile phones, operation	9	5
Fixed networks, operation	72	54
Cordless phones, operation	22	13
Broadband modems and routers, operation	35	21
Private branch exchanges, facsimile machines and various business systems, operation	35	20
Personal computers, operation	258	155
Data centers, enterprise networks, and transport networks	226	170
Data centers, operation	180	108
Enterprise networks, operation	29	17
Transport networks, operation	17	10
TVs, operation	340	204
TV peripherals, operation	136	82
TV networks, operation	30	22
Other E&M devices, operation	70	42
Total	**1,509**	**969**

Note: Data sourced from Malmodin and co-workers (2010, tables 4 and 5). The calculations in table 6.1 exclude manufacturing of the devices.

kilometers in length, or enough to wrap around Earth's equator twenty-five times (see figures 6.15 and 6.16 online). Undersea cables are engineered for service lives on the order of twenty-five years, but actual use times vary. When the cost of running data across a cable exceeds the cost of sending the same amount of information over newer, higher-capacity lines, cables become "economically obsolete" and are usually abandoned in place (TeleGeography 2017). Humans have been stretching wires across oceans for almost 160 years (Finn and Yang 2009), and the greatest part of that economically obsolete infrastructure remains where it was laid. The materials of which undersea cables are composed are thought to be inert in seawater (though how their materials, such as the polyethelene sheathing of newer cables, interact with the ocean environment is not known; see United Nations Environmental Programme 2009). In this sense, their environmental impacts are thought to be minimal or benign. This infrastructure is also colonized by marine life, and some experiments in using abandoned cables as anthropogenic reefs have been successful on the eastern seaboard (Starosielski 2015), though there are open questions about whether such reefs are a net benefit for marine life (Seaman 2007; Van der Stap, Coolen, and Lindeboom 2016). What remains is the shear mass of retired cable on the seabed—and this is increasingly being explored as a source of recyclable materials. CRS Holland, a Dutch cable recovery venture, estimates that 94 percent of retired cables remain on the ocean bottom. The firm calculates this amounts to 40,000 metric tons of aluminum, 600,000 tons of copper, 900,000 tons of polyethylene, and 1.5 million tons of steel, together worth more than U.S. $4 billion (CRS Holland 2017).

The discardscapes of clicking are generated in all Earth environments—atmospheric, marine, and terrestrial. They can also be charted in extraterrestrial space around the planet. Historian of science and technology Lisa Ruth Rand highlights what she calls "the omnipresence of GPS" as a core technology for organizing modern life (Rand 2014). GPS relies on some twenty-four satellites orbiting 20,350 kilometers above Earth. The signals they produce are crucial for timing, location, and synchronization applications across a vast array of digitally enabled infrastructure. Networked communication infrastructure such as cell phones requires GPS to coordinate data transmission. Financial transactions, including everything from stock market deals to ATM cash withdrawals, depend on GPS to coordinate and validate flows of money while also preventing fraud. Transportation systems from cars to trains to boats and airplanes rely on GPS for navigation and scheduling. Other suites of satellites provide remote sensing data that feed weather reports into our digital devices. News reports travel via satellite to broadcasters, who push out reports over terrestrial and marine-based Internet infrastructure. The discardscapes of these and other space-based technologies have been accumulating in extraterrestrial space since the launch of Sputnik in 1957 (Parikka 2015; Parks 2012; see figure 6.17 online). Current official estimates put the mass of spent rocket bodies and debris in Earth orbit at more than 3,400 metric tons, or just shy of 46 percent of the mass of all objects, including functional spacecraft currently in orbit (NASA Orbital Debris Program Office 2017, 13). In other words, nearly half of what orbits Earth is a discardscape of human technological action when measured by weight (Gorman 2014; Lepawsky 2016).

Conclusion

The familiar framing of e-waste as a postconsumption waste management problem is peculiar. That framing brackets out the vast quantity of waste arising ubiquitously but unevenly at all points in the existence of electronics, but especially the waste arising from resource extraction and manufacturing. This situation points to why problem definition matters so much when it comes to waste issues. If we accept the customary way of framing discards from electronics as a postconsumer waste problem, then we guarantee that solutions so framed will not match the locations or the scales of the most significant sources of waste (Liboiron 2014). Even if complete consumer compliance with electronics recycling programs were possible, most waste arising from electronics would go unmitigated.

The usual way of framing e-waste needs to be made strange again. It needs to be defamiliarized. This chapter has gone some way toward doing so. I have charted what I called the discardscapes of electronics into Earth environments, human bodies, and extraterrestrial space. From a single site of copper extraction in the archipelago of minescapes came more waste than the total annual tonnage of postconsumer e-waste arising in the United States. The amount of waste acid arising annually at a single smelter facility is more than double the highest estimates of tonnage of e-waste exports from the United States. A single tailings pond spill over a twenty-four-hour period wipes out drinking water on which 25,000 people depend. Production technologies are putting chemicals into the atmosphere that will remain there for centuries to millennia. Those chemicals are thousands of times more potent than CO_2 in terms of their global warming potential. At the same time, the manufacturing of electronics leaves plumes of toxicants in groundwater and in workers' bodies. Cleaning up these wastes arising in manufacturing leads to its own discards. For every pound of contaminant extracted for remediation from Silicon Valley's former electronics manufacturing zones, about 4,000 pounds of CO_2 are released into the atmosphere from those treatment processes. Clickscapes demand materials and energy. From those demands can arise the methylmercury poisoning of indigenous food webs. Enough discarded cabling to circle the Earth multiple times sits on the ocean floor. The debris of satellites organizing today's reliance on GPS for financial transactions and global shipping orbits the Earth. Some of it will do so for centuries to come.

If the discardscapes charted in this chapter, however incompletely, are to be mitigated then approaches other than a reliance on postconsumer recycling must be demanded and instituted. I provide some outlines of such approaches as I draw the book to a close in the next and final chapter.

7 Conclusion

Any system that deals with the four aspects of modern waste—tonnage, toxicity, heterogeneity, and externalization—on a large scale will change waste infrastructure and what counts as trash.
—Max Liboiron (2013a, 12)

Introduction

As the fiscal year closed in April 2017, Apple released its new Environmental Sustainability Report, covering fiscal year 2016. What grabbed the most media attention was the company's stated ambition of "challenging ourselves to one day end our reliance on mining altogether" and make Apple's products entirely from recycled materials (Apple 2017, 16). Apple's goal, while laudable, also embodies a more general paradox. Assuming the firm one day achieves its goal, it will have lessened or even eliminated its demand for the mining of the raw material resources needed to make its products. Yet the aggregate demand for those resources may not be affected at all or may even increase. The metals once mined to satisfy Apple's manufacturing needs will, after Apple exits that market, become available to other firms (whether in the electronics sector or not) to use in their own manufacturing processes. While Apple's demand for such resources might be diminished, aggregate demand might not decline. Wittingly or not, Apple's strategy springs what has been called the recycling trap (Turner 2007, 208). That trap entails deferring or forgoing innovations in product durability, repairability, and reusability that would extend the working life of equipment in favor of product destruction for materials recovery— otherwise known as recycling. It may also involve treating as separate issues making changes that would ameliorate the use of toxic substances in the manufacturing process or the products themselves, even though those substances enter bodies and environments as by-products of the manufacturing process.

Solutions to the wastes arising ubiquitously but unevenly at all points in the production and life of electronics will have to go well beyond a reliance on recycling. In what follows, I sketch in broad outline some concrete measures that could be pursued to reduce the amount of waste associated with electronics. Such measures include decriminalizing export, facilitating ethical trade, implementing genuine extended producer responsibility, and democratizing the industry. Ultimately, however, it is necessary to achieve a radical aggregate reduction in the material and energy throughputs of the electronics sector and its resulting products.

Decriminalize Export for Reuse, Repair, and Elective Upgrade

One way to mitigate waste from electronics is to decriminalize export for reuse, repair, and elective upgrade (e.g., swapping a black-and-white TV for a color TV made out of a CRT monitor originally intended for a desktop computer. On elective upgrade more generally, see Ingenthron 2012). The benefits of exports for such purposes are many. Reuse, repair, and elective upgrade all conserve the energy and materials embodied in devices during the mining of resources for them and the manufacturing of them. Chapter 6 discussed the weighty geographies of minescapes and productionscapes. It showed that the discards and harms arising from the mining of materials and the manufacturing of electronics vastly exceed those that arise from consumers discarding their devices. Reusing, repairing, and upgrading electronics are ways to conserve the energy and materials expended to make them in the first place.

Other benefits of export for reuse, repair, and elective upgrade are socioeconomic. They include enhancing the ability of larger numbers of people to access electronic devices—and the attendant ability to get online—than would otherwise be possible. The UN special rapporteur on the promotion and protection of the right to freedom of opinion and expression concluded in 2011 that "the Internet has become an indispensable tool for realizing a range of human rights, combating inequality, and accelerating development and human progress," and urged all governments to "make the Internet widely available, accessible and affordable to all segments of the population" (La Rue and United Nations 2011, 22). Export for reuse, repair, and elective upgrade is crucial to achieving such benefits since it is not with the newest version of a smart phone, tablet, laptop, or other digital device that most of the world can (or will) get online (Giridharadas 2010).

Critics of this argument for export as a bridge over the digital divide claim that such exports are really just pollution externalization in disguise. Yet cases like Joseph Benson's, discussed in chapter 3, suggest otherwise. Certainly the shipments he was convicted of making met a legal definition of waste, but even that definition acknowledged such discarded equipment could be economically reusable. The economics of export for disposal cost avoidance (i.e., pollution abatement cost externalization) did not explain Benson's motivations in the case since he and his colleagues were buying discarded equipment at civic amenity sites and at the same prices required for legally proper recycling. That equipment may not have been economically repairable in the UK, but no determination was made about the economic viability of such repair in the countries to which it was headed.

While Benson's case may be just one example, a recent in-depth analysis of the illegal trade in discarded electronics out of Europe found that "the main driver behind exports is the reuse value combined with the avoided costs of sorting, testing and packaging" (Huisman et al. 2015, 18). This finding is not a ringing endorsement of the export trade, to be sure, but neither does it constitute decisive support for the prevailing explanation of exports of discarded electronics being about externalizing pollution abatement costs. The same report notes that "the magnitude of the reuse value" of mixed, unsorted, and untested equipment sent to Africa "is multiple times the material value" (Huisman et al. 2015, 18). Of the 1.5 million tons of discarded electronics leaving the EU, 200,000 tons are properly documented and perfectly legal to export. The remaining 1.3 million tons, however, exist in "a grey area subject to different legal interpretations and susceptible to export ban violations" (Huisman et al. 2015, 16). A zone comprised of 86 percent of the equipment in question is a rather large "gray area" of uncertainty and points to the need for much-improved regulations.

It bears remembering that the Basel Convention already permits exports of discarded electronics under the stipulations of Annex IX. As long as the exports are for reuse that includes "repair, refurbishment or upgrading, but not major reassembly" (Basel Secretariat 2011a, 83), they are permissible under the Basel Convention. Of course, as I noted in chapters 2 and 3, the convention's silence on what counts as major reassembly leaves much room for highly polarized debate as to the permissible conditions for such exports.

Pursuing the decriminalization of exports for reuse, repair, and elective upgrading will likely put in jeopardy ongoing efforts to prohibit exports based on a fixed geography of sending and recipient countries. As discussed in chapter 2, the Ban Amendment of the Basel Convention seeks to ban exports originating in the EU or the OECD to any other signatory country not part of those regional blocs. Yet, as I showed in chapter 4, the geography of trade has undergone a shift that renders Annex VII's geographically based prohibitions against exports increasingly irrelevant. The largest total volume of postconsumer e-waste already arises domestically in Asia (Baldé et al. 2015; see also Lepawsky 2015a, 2015b; Lepawsky and Connolly 2016). As more and more countries of the global south (i.e., non–Annex VII states) become consumers of electronics, the more postconsumer e-waste will arise in countries located in the region and the more those discards will be traded between geographies that do not match export prohibitions against Annex VII to non–Annex VII trade.

This ongoing change in the geographies of trade may mean it is worthwhile to rethink whether the nation-state is the best scale at which to pursue decriminalization efforts. Regional blocs such as the EU, the signatories to NAFTA, or the Association of Southeast Asian Nations (ASEAN) may be far more relevant. Other possible geographies of regulatory reform to decriminalize export for reuse, repair, and elective upgrade exist. For example, since it is now technically feasible to surreptitiously track individual pieces of equipment with submeter precision, one could imagine using such tracking techniques to regulate exports not to this or that whole country but to a specific facility or individual operation. Operations receiving exports for reuse, repair, and elective upgrade could be licensed or certified for such activities so that they conform to the Annex IX meanings of repair, refurbishment, or upgrading—if indeed such licensing or certification was seen as desirable by those most deeply affected by it.

Certified electronics recycling systems have emerged as one way to facilitate such trade. Two of the most prominent systems are e-Stewards, an initiative of the nonprofit Basel Action Network (BAN), and R2, run by another nonprofit called Sustainable Electronics Recycling International (SERI). Broadly speaking, both certifications intend to achieve the same thing: verify that discarded electronics are managed in ways that are safe for people and environments. Both systems permit exports for processing if all facilities throughout the export chain are certified to each system's standards. One important difference, however, is how the two systems handle exports when repair is involved. E-Stewards requires that all equipment be fully functional before exports for reuse are permitted (E-Stewards and Basel Action Network 2013, 6). In contrast, R2 allows exports under the category "R2 Ready for Repair," which does not require the equipment in question to be fully functional (SERI 2014, 11). Deep divisions between the two organizations exist on these different ways of handling repair (for more in-depth discussion, see Lepawsky 2012, 2017).

Tracking, licensing, and certification are possible, but not necessarily desirable, approaches to decriminalization. Despite the entrenched divisions between the two certification systems, they share a willful bracketing out of any consideration for the roles that might be—or are already—played by actors in the informal sector. Costs to become certified by either system are vastly out of reach for any actors in the informal sector, irrespective of whether they operate inside or outside the OECD or EU. More pointedly, both certification systems at least implicitly see the informal sector as a competitor that robs material that could otherwise enter the certified systems—and the revenue streams they generate for BAN and SERI. The reasons behind this bracketing out need not be interpreted as inherently cynical or economistic ploys. At least part of the problem stems from how the informal sector is typically dismissed as subordinate to, or parasitic on, the formal sector (for helpful introductions to general issues of informality, see Chen 2012; Meagher 2013; on the critical role of the informal sector for urban recycling, see, e.g., Chaturvedi 2014; Gill 2009; Gutberlet 2008). Specifically within e-waste debates the informal sector is often treated as a monolith defined by one link—those engaged in "backyard" or "primitive" recycling—in an otherwise highly diverse and skilled network of economic action. Of course, environmentally harmful practices occur at some points in this network. But those points are not the network. Skilled yet informal repair for reuse and elective upgrade makes major contributions to the local economies where it happens (Gu et al. 2016;

Lepawsky and Billah 2011; Reddy 2015) while also conserving the energy and materials embodied in used electronics.

Other socioeconomic benefits include employment in the vast yet largely underresearched formal and informal economies supported by export for reuse, repair, and elective upgrade. Economic activity in this sector is highly diverse in terms of firm size and revenue, depending on how one classifies it. It can cover everything from large international brand names that outsource repair to contracted facilities in Asia to micro, small, and medium-sized enterprises in both the global north and the global south. Examples of the latter come from my research experience in Dhaka, Bangladesh. In Dhaka some 1,500 independent reuse and repair businesses were found to cluster along a 1.5-kilometer section of road in the city (Lepawsky and Billah 2011). These and other businesses like them depend heavily on an influx of discarded but reusable and repairable electronics imported from other markets in the region such as China, Singapore, South Korea, and Taiwan (Lepawsky and Connolly 2016). Many of these businesses are informal operations, but informality is not confined to the global south, as recent work from Spain on the recuperation of value from discarded electronics in Barcelona shows (Callén and Criado 2016). In the formal sector of the global north, iFixit is only the most prominent example of a substantial infrastructure of independent electronics repair. The company claims its free online repair manuals were accessed by 94 million people around the world in 2016 (Whitford 2017). Mapping the environmental and socioeconomic effects, both positive and negative, of the formal and informal information and communications technology (ICT) repair sector remains an important task for future research.

Facilitate Ethical Trade in Electronics Reuse, Repair, Refurbishment, and Recycling

Even the harshest critics of export have conceded that electronic medical devices deserve special consideration in the matter of legalizing export for reuse, repair, and elective upgrade, as I noted in chapter 2. It is possible, then, to enact solutions that go beyond merely decriminalizing export for reuse, repair, and elective upgrade. In broad terms, such solutions would entail facilitating bidirectional trade (both exports and imports) within and between the OECD, EU, and locations outside those regions. Other models already exist for facilitating exports of discarded electronics to markets outside the OECD and the EU in exchange for locally collected end-of-life electronics in those markets, which are then sent for recycling in facilities back in the OECD or EU. One such model, called "Best of 2 Worlds" (Bo2W), has been developed under the aegis of the UN (Manhart 2010; Wang et al. 2012). Two pilot versions of Bo2W have been tried. One involves collecting discarded electronics in markets outside the OECD and EU, performing local manual disassembly under high standards of occupational health and safety for workers, and then exporting the resulting material fractions to state-of-the-art processing facilities in OECD or EU locales. The second version of the model is identical except that it also includes the collection of discarded electronics in the OECD and EU, followed by their export to markets outside those regions for manual disassembly, followed by reimport of the resulting material fractions back to OECD and EU processing facilities. Advocates of these models argue that two broad benefits are achieved from them. First, workers in the disassembly phase gain access to much safer working conditions and higher wages than they would typically experience in the informal sector. Second, OECD and EU facilities gain access to material feedstocks of much higher purity than those that result from the automated shredding of electronics (automated shredding creates material mixing problems that must be sorted out using expensive machinery; material fractions resulting from those processes are typically higher in impurities than those that result from manual disassembly). The overall benefit of Bo2W models, advocates argue, is a reduced reliance on traditional resource extraction mining and the increased reuse of materials already extracted.

Critics have identified a number of drawbacks to the Bo2W approach that raise very important questions about whether it can lead to unreservedly positive outcomes. The geographer Rajyashree Reddy's ethnographic research (2013, 2015, 2016) on the implementation of Bo2W in India shows that it relies on public subsidies to private recycling firms so that they may compete against the informal sector. The results include enhanced precarity for informal workers pushed out by the formalized Bo2W system. Freyja Knapp (2016) situates Bo2W within broader shifts in the political economy of mining, in which "the urban" is a new kind of resource frontier. As such, she asks readers to consider the socioeconomic justice implications resulting from the mining of waste material arising in already impoverished cities so that high-quality feedstock can then be exported to facilities headquartered in much richer countries. As colleagues and I have argued elsewhere, another concern with the Bo2W approach is how it entrenches a system of collect-and-destroy materials recovery (i.e., recycling) with no attention at all to the resource conservation gains to be realized from the reuse, repair, or elective upgrade of digital equipment (see Lepawsky et al. 2017 for further discussion).

Another model is provided by the fair trade electronics recycling efforts of the World Reuse, Repair and Recycling Association (WR3A). This model includes paying discount prices for imports of quality reusable, repairable, and upgradable equipment to offset the cost of collecting equipment that has reached its locally relevant end of life in export markets. While used electronic equipment often has a much longer service life in the global south, no such equipment will continue in use forever. Discount pricing on shipments of quality used electronics can be offered under contracts that require the importer to return to the exporter shipments of locally collected end-of-life electronics for recycling at facilities that do not exist in the local market (see Ingenthron 2017a, 2017b, 2017c). The organization points out that while the export trade is not perfect, "tinkering, repair, refurbishing and reuse have created affordable cell phone and internet systems across the globe" in markets that would otherwise be much diminished in access to ICT because of the costs of new equipment relative to local incomes (Ingenthron 2017a). The WR3A has developed a contract-based model that relies on the premise that high-quality reusable and repairable equipment can be exported to markets outside the OECD and the EU. The contracts stipulate that importers agree to send back a mass balance of end-of-life electronics locally collected from the export market to be dealt with in state-of-the-art recycling facilities in OECD or EU markets. The costs of sending the end-of-life equipment back to OECD or EU facilities are covered by the price premium achieved from reuse markets in the export destination.

A similar model has been developed by WorldLoop, an organization dedicated "to extending the positive impact of ICT projects in developing countries by offsetting the negative environmental impact of its hardware" (WorldLoop 2013). To achieve this, WorldLoop supports initiatives owned and managed outside the OECD or EU to safely collect, dismantle, and recycle locally discarded electronics. WorldLoop facilitates exports of those fractions to state-of-the-art facilities in Europe when no facilities exist in the local collection markets for the safe disposal of hazardous fractions from this equipment.

Institute Genuine Extended Producer Responsibility

The Canadian Council of Ministers for the Environment, a consortium of provincial authorities, outlined a vision for implementing extended producer responsibility (EPR) for electronics in 2004 (Canadian Council of Ministers of the Environment 2004). In that document, EPR is defined as environmentally sound management consistent with "the 4R waste management hierarchy [of] reduce, reuse, recycle, recovery" (Canadian Council of Ministers of the Environment 2004, 1). In broad terms, EPR systems are intended to shift the costs of managing end-of-life equipment away from public authorities and general taxpayers and toward brand manufacturers. The goal is to use EPR-based systems to send a price signal to manufacturers and thereby financially incentivize the design and manufacture of devices that would ultimately reduce the costs of handling them at end-of life. Such cost reduction could be reached in more than one way. For example, brand manufacturers receiving the signal that they will have to pick up the costs of take-back might find ways to save on those costs by making their devices easier to recycle. Other options are possible, such as making devices that are more durable or more repairable.

While goals like these are what it is hoped EPR will achieve, the way EPR legislation typically sets up the financing regime for a given system militates against actually achieving EPR. For example, in Canada, the United States, and throughout the EU, EPR systems for electronics take-back exist (on the EU system, see Sander et al. 2007; see Lepawsky 2012 on EPR for electronics in Canada and the United States). Though their details differ, they share the common characteristic of being financed by consumers at the time of original purchase. When consumers buy a new device, they are assessed a fee to cover costs incurred to collect and recycle those devices at some future point. Depending on the specific EPR regime in place, these fees may or may not be visible to consumers. Some jurisdictions require purchase price information to include a separate indication of the EPR fee being charged (thus making the fee visible). Other jurisdictions allow the fee to be wrapped into the total purchase price of a given device, thus making the fee invisible.

A number of implications result from either approach. For example, it is consumers, not producers, who end up financing the take-back infrastructure. This in itself is not an inherently negative result. One could argue that consumers of a particular class of commodities should bear a specific economic responsibility for covering the costs of managing those devices when they are discarded. On the other hand, installing EPR systems financed by consumers guarantees that producers feel no price signal that might otherwise incentivize them toward clean(er) or green(er) design and manufacturing. In other words, EPR systems financed by consumers ensure that producers continue to be able to externalize the costs of managing discarded equipment. Consumers sometimes interpret the take-back fees as an instance of government taxes even though it is usually a nonprofit or industry-run organization that handles the fees collected for take-back. This can have the effect of deflecting public irritation over increased prices from manufacturers to governments.

A genuine EPR system would enable extending the useful life of devices by, for example, incentivizing durability, reusability, and repairability. Yet where EPR systems for electronics exist, the opposite has been achieved. Despite the existence of EPR legislation in Canada, the United States, and the EU over the past decade, it is not unusual for new models of devices to be less durable and less repairable than the models they replace. At the same time, brand makers increasingly turn to recycling as a preferred method of handling discarded equipment. As I explored in chapter 6, postconsumer recycling will not recoup the material and energy discards that arise from the two most significant sources of waste from electronics, resource extraction and manufacturing (more generally, see Liboiron 2009; MacBride 2012). Apple Computers has come under particular scrutiny for its "must shred, no reuse" policies built into its agreements with third-party recyclers of its products (Koebler 2017). These policies guarantee that all energy and material conservation gains to be gleaned from extending the useful life of devices are lost. What drives such policies is largely liability concerns over data security and brand identity. While not illegitimate concerns outright, they tell us that priorities other than resource conservation hold sway, even though EPR legislation has been instituted for at least a decade.

Critics of genuine EPR might argue that costs will always be passed on by brand manufacturers to the consumer. But why should this necessarily be so? Regulatory action could be taken to prevent this. An example would be a requirement that manufacturers use some portion of their end-of-year profits to finance take-back. Focusing on profit would increase the chances that brand manufacturers would find ways to innovate so as to save costs at time of take-back. EPR regulation of this type would, then, sculpt genuinely novel approaches to reducing the tonnage, toxicity, and heterogeneity of e-waste while giving brand manufacturers the leeway to achieve those reductions by whatever means of innovation they can to achieve them.

Demanding that some increment of profit be used to finance such innovation is not such a radical idea. Existing initiatives such as 1% for the Planet, which facilitates private sector donations to various environmental projects, takes a similar approach (onepercentfortheplanet.org 2017). It provides a platform for businesses to commit 1 percent of the value of total sales of a whole organization, a specific brand, or a product line to a variety of philanthropic projects.

There will be critics who dismiss this kind of initiative as what Peter Dauvergne (2016) has called "environmentalism of the rich." They would be correct to do so. With its focus on a percentage of total sales, 1% for the Planet does nothing to slow aggregate growth in resource extraction, manufacturing, or consumption or the waste arising from any of those activities. Little, if any, reduction in the tonnage, toxicity, or heterogeneity of wastes will occur until the costs associated with mitigating each of them are internalized by firms at all points of the existence of electronics. This is why genuine EPR must enforce internalization of those costs if actual reductions in waste and changes in what counts as waste are to be achieved.

Yet genuine EPR on its own is not enough. As Zsuzsa Gille (2007) argues, production itself must be much more radically politicized. If politicizing production sounds strange or threatening, I would reply by pointing out that production is already politicized in many ways—even if it may not seem so because that politicization is so axiomatic it typically disappears from view. For example, chapter 6 discussed data from the EPA's Toxics Release Inventory on how much, of what type, and where such releases are continuously occurring from the computer and electronics sector.

Facilities are required by law to report such data because they concern waste deemed hazardous. No similar legal requirement to report releases of nonhazardous waste exists. This is a politics of measurement, as discussed in chapter 5. It is a situation in which one thing, hazardous waste, is measured while another, waste, is not measured at all (for further discussion, see MacBride 2012). As a consequence, U.S. citizens can have a reasonably good idea about how much hazardous waste is released by production facilities but little or no idea about how much other waste arises from those facilities. In this sense, production is already politicized.

To return to the question of EPR specifically, its implementation too politicizes production in some ways but not others. For example, if we agree that consumers alone should finance EPR systems we are also, in effect, consenting to the idea that public decision making should be limited to what to do with waste after it already exists. We are simultaneously consenting to the idea that those in the best position to effect changes that would reduce or eliminate wastes before they arise—manufacturers—should be free to socialize the costs of waste arising in design and manufacturing while privatizing the profits that result. So a call to politicize production is not itself radical, since production already is politicized in some ways. Instead, it is a question of how production is politicized that matters. If politics is that which pertains to struggles over various forms of power, then the divisions of power that EPR implements amount to collective struggles over the right and proper distribution of public and private abilities to act in certain ways. These are struggles that might benefit from enhancing the democratization of industry.

Democratize Industry

What might it mean to democratize the electronics industry? Broadly speaking, it would involve instituting forms of public oversight of electronics production that reach beyond the factory gate and into the practices that enact the tonnage, toxicity, heterogeneity, and harms of waste that will eventually arise at all points in the existence of electronics. This would mean, among things, instituting forms of public decision making in design and manufacturing. Again, it is Zsuzsa Gille who leads me to this point. She asks what the value of democratic institutions for making environmental claims is "if their scope of authority is limited to policies that can,

at the most, remedy but not prevent waste problems? If waste production—
that is, how much and what kind of wastes can be produced—is excluded
from public discourse, the most that democracy can achieve is to regulate
what to do with the wastes already produced" (Gille 2007, 210; see also
Thorpe 1992). At a minimum, Gille argues, citizens need access to data
about wastes arising upstream in resource extraction and manufacturing.
They also need to have actual deliberative input into questions about how
much and what kinds of materials should and should not be produced
at all.

The EU's Restriction of the Use of Certain Hazardous Substances (RoHS)
and Registration, Evaluation, Authorisation and Restriction of Chemicals
(REACH) regulations are two pieces of legislation that go some distance
toward the democratization of industry in the sense of enhancing public
oversight of design and manufacturing. To some extent, that oversight also
affects how much and what kinds of waste can be produced. RoHS went
into effect in 2006 and REACH followed in 2007, with a phased implementation approach. The two pieces of legislation are quite different in the
ends they seek to achieve but are similar in that they change the tonnage
and toxicity of materials that will eventually become waste. RoHS is
directed specifically at the electronics sector. It restricts six substances—
lead, mercury, cadmium, hexavalent chromium, polybrominated biphenyls (PBBs), and polybrominated diphenyl ethers (PBDEs)—to maximum
allowable concentrations by weight of materials used in manufacturing.
REACH, in contrast, is far broader in scope. It applies to all manufacturers
and importers regardless of sector and seeks to mitigate the health and
environmental risks of all chemicals used in the EU.

Broadly speaking, REACH seeks to regulate the most hazardous substances out of production by mandating the use of less harmful or non-
harmful substitutes (European Chemicals Agency 2017). Among other
things, REACH requires that data on chemicals used be communicated
throughout the supply chains of a given product. Companies that use
chemicals must communicate to their suppliers what the chemicals are
used for, how workers and consumers use their products, and how those
chemicals enter the environment. They must also actively seek out safe(r)
alternatives and use them where possible. Meanwhile companies that
manufacture chemicals must register with the European Chemicals Agency.
They must also communicate to their customers how the chemicals they

produce can be used safely, and they must actively develop and market safe(r) alternatives.

While neither RoHS nor REACH is specifically about managing waste, both are nevertheless highly significant forms of public oversight that will affect waste arising in manufacturing. RoHS and REACH get beyond the factory gate and into design and manufacturing processes themselves so as to affect the chemical composition of commodities before they are manufactured. This is crucial to achieving reductions in tonnage and toxicity arising in manufacturing before that which will become waste is made in the first place. Moreover, though these are both pieces of legislation enacted in the EU, they become de facto global standards because of the size and wealth of the EU market. Electronics manufacturers will conform to RoHS and REACH requirements if they wish to sell into the EU market, and it will make little economic sense to create product lines that are chemically different if they are also sold into other markets in the OECD or elsewhere.

In thinking through other concrete proposals for democratizing the electronics industry, it helps to keep in mind an aphorism attributed to the heterodox economist Kenneth Boulding: "If it exists, then it is possible" (quoted in Chapman 2006, 1471). Three multibillion-dollar sectors—food, pharmaceuticals, and automobiles—provide useful lessons for imagining a more democratic and safer material world of electronics.

Lessons from the Food, Pharmaceutical, and Automobile Sectors

Today, consumers in the United States (and similarly in Canada and Europe) take it for granted that the foods they eat and the medicines they take are generally safe for consumption. The extension of public institutional requirements into the design and manufacturing phases of food and pharmaceuticals makes this possible. Public institutions such as the U.S. Food and Drug Administration, Health Canada, and the European Medicines Agency set standards that private manufacturers must meet or exceed if their products are to be sold to consumers. Broadly speaking, food and pharmaceutical manufacturers must demonstrate that products are chemically safe for consumption within mandated ranges of risk. In effect, manufacturers in these sectors are required to demonstrate the safety of their products *before* those products are put on the market. An important if unintended consequence is regulation over the material composition of that which will eventually become waste. It is also a regulatory approach of relatively recent vintage. Key amendments to the U.S. Federal Food, Drug, and Cosmetic Act, for example, occurred only in 1962, partly in response to the thalidomide tragedy in Europe. Those changes increased the rigors of public oversight over drug safety in several ways, including by increasing the weight of evidence of safety drug manufacturers must reach before approval by the FDA, by requiring that manufacturing facilities be regularly inspected, and by requiring manufacturers to register annually with the agency (U.S. Food and Drug Administration 2002, 22).

Similarly, today it is expected that automobiles will come with safety features such as seat belts, laminated glass, and brake systems that meet or exceed regulatory standards. A little over forty years ago no regulatory requirements mandated such features. Car companies adhered to voluntary standards created by an industry organization, the Society of Automobile Engineers (SAE), first headed by the car manufacturers Henry Ford and Andrew Riker. The SAE reigned for some sixty years before the organized consumer activism of the Consumers Union in the mid-1960s eventually led to the creation of the National Highway Traffic Safety Administration (NHTSA). The creation of NHTSA wrested jurisdiction over safety away from the voluntary industry standards body of the SAE—which was supported by the industry oligopoly of GM, Ford, and Chrysler—and placed it in public hands. The NHTSA has jurisdiction over motor vehicles in the United States and prescribes safety standards cars must meet before they can be sold on the market.

The point I am making here is not about the specifics of the U.S. FDA, the NHTSA, or the historical details of these particular institutions (for such details see Flink 1990; Thomas 2014). Nor is my point that the FDA or NHTSA represents some kind of regulatory template that could or should be imposed on the electronics sector. Instead, I wish to craft an argument for the democratization of the electronics industry by way of analogy with these institutions. We accept as normal—now—that multibillion-dollar industries—food, pharmaceuticals, automobiles—are subject to substantial public oversight of the safety of the products they create *before* those products are put on the market for sale. That sense of what is normal and taken for granted is of quite recent origin. But if such norms exist with respect to some multibillion-dollar sectors of the economy, then it is possible to enact them into existence for other sectors, such as electronics manufacturing. The electronics sector could be made subject to similar public oversight that would affect the material safety of its products before they can be sold. The history of organized consumer activism that led to the creation of the FDA and the NHTSA tells us we could in principle world electronics otherwise, even if it might not be easy to do so. The electronics sector is, of course, already subject to all kinds of public oversight, such as occupational health and safety legislation, electrical safety standards, and the like. My point is certainly not that the electronics sector is somehow regulation free. However, there is little public oversight over such matters as how much and what types of waste that arise at all points in the existence of electronics are or should be permissible. So if it initially sounds exotic to suggest democratizing the electronics industry, we should keep in mind that democratization in some form has already happened in other sectors. And if it exists, it is possible.

The automobile industry also offers lessons about expectations concerning the repairability of products sold to consumers. In his rich history of technological change, expertise, and the shifting social status of auto mechanics in the United States, Kevin L. Borg charts what he calls technology's middle ground—a space "between production and consumption in which workers maintain and repair artifacts that they neither create nor own" (Borg 2007, 2; see also Lucsko 2016). In the automobile sector this middle ground has for decades been populated by a vast archipelago of brand, independent, and franchised aftermarket maintenance and repair businesses. The sector employed more than 374,000 people in the United States in 2012 and had total revenues exceeding $45 billion (U.S. Census Bureau 2012). Much as with the chemical safety of products, consumers expect that their cars can be maintained and repaired so as to extend the operational life of these machines. Repairability has been a norm in this multibillion-dollar sector for some time, notwithstanding what some interpret as maneuvers by manufacturers to wrest control of maintenance and repair out of the hands of both owners and the independent sector (repair .org 2016).

Advocates of "right-to-repair" laws are seeing some legislative successes in the United States, but winning the right to repair will be a hollow victory if repairability itself is engineered out of products. The contemporary design of cell phones, laptop computers, and desktop computers often (not always) results in declines in the repairability of newer models (this is especially the case when firms substitute adhesives for screws to bind devices or key components together). When such machines break, there is sometimes no—or no easy or affordable—fix that will extend their operational life. When that happens, and if those machines are sent for recycling, the conservation of energy and materials expended in resource extraction for, and manufacturing of, those machines is lost. As discussed in chapter 6, no amount of postconsumer recycling can recoup the energy and materials already expended and discarded upstream in resource extraction and manufacturing.

The net environmental effects of recent advocacy around consumers' "right to repair" products they purchase, including electronics, is still not clear. Whether those products are repaired by individual consumers, independent repair shops, or brand manufacturers makes little difference in terms of conservation of resources. Repair for reuse that leads to conservation of resources reduces the environmental footprint of those devices, regardless of who does the repair. If, on the other hand, consumer organizing results in brand manufacturers making more durable and repairable devices, then right-to-repair advocacy could lead to at least some net environmental gains. Such gains may be fleeting, but the broader lesson bears repeating: if it exists, it is possible. The repairability of automobiles is a norm. It could be made so in the electronics sector as well.

Degrow

Ameliorating the tonnage, toxicity, heterogeneity, externalization, and harm of wastes arising from electronics is ultimately dependent on achieving aggregate reductions in each of those characteristics at all points in the existence of electronics. Two broad parallel processes are necessary to achieve this desired scenario: an absolute reduction in or elimination of toxicants from electronics and, equally important, an aggregate reduction in the total throughput of matter and energy associated with the existence of electronics. Without radically reducing or eliminating toxicants, even if reducing total throughput of matter and energy is achieved, some toxicants will continue to accumulate ubiquitously but unevenly in bodies and environments. This is because, as I noted in chapter 6, some chemical releases currently associated with manufacturing electronics are effectively permanent additions to the Earth. Other chemical constituents of the devices themselves, such as those found in the plastics that compose them, constitute a class of pollutants known as persistent organic pollutants (POPs). Depending on specific chemical constituencies, POPs can also be effectively permanent. As a consequence, unless such pollutants are eliminated altogether, even radically reduced aggregate demand will still lead to their growing accumulation in Earth environments (Gray-Cosgrove, Liboiron, and Lepawsky 2015).

Referring to a need for aggregate reduction in material-energy throughput is one way of describing what others have called "degrowth" (D'Alisa, Demaria, and Kallis 2014). Degrowth is a translation of the French *décroissance*, which began appearing in the literature on economics in the early 1970s, partly owing to the influence of the expatriate Romanian statistician and economist Nicholas Georgescu-Roegen (1970). Translation is always an imperfect enterprise, and what degrowth means today remains up for debate. However, two broad threads can be traced in the burgeoning degrowth literature: questioning the pursuit of economic growth as a self-evidently positive end in itself and the articulation of desires toward convivial sufficiency. There is more than one way to achieve such desires, but doing so requires fundamental shifts in how we organize economic action. It bears keeping in mind that the idea of economic growth as an unalloyed good that should be explicitly supported by public policy is actually a very recent invention, coming to the fore only in the mid-twentieth century (Schmelzer 2016; Victor 2008). With respect to electronics specifically, degrowth would entail a radical reorientation of the norms of resource extraction, manufacturing, and use of these technologies.

To understand why degrowth in some form or other is necessary if waste from electronics is actually to be reduced, it is important to think with a concept some economists call the Jevons paradox or the rebound effect. William Stanley Jevons was a nineteenth-century political economist who studied the relationship between changes in the efficiency of coal use and aggregate demand for coal. Jevons's findings pointed to a fundamental contradiction in the burgeoning industrial use of coal: as new techniques were adopted that increased the efficiency with which coal was used, aggregate demand for the resource increased rather than decreased. In other words, the more output per unit of coal that could be obtained by using more efficient techniques, the greater the demand for coal became. Instead of lessening the demand for coal, efficiency gains led to a growth in demand for the mineral. Why? If less coal is needed per unit of output, more coal is available on the market, cheapening the price of coal. Those who used less coal—or none at all—before prices fell can now afford to use more. As a consequence, overall demand rises; that is, demand "rebounds" after efficiency gains are made. This is a somewhat simplistic explanation of the Jevons paradox, but it captures in broad terms a major area of research by contemporary ecological economists relevant to the specific issue of waste arising from electronics.

Contemporary studies of the rebound effect produce results with wide margins of uncertainty about what if any aggregate resource conservation gains are made through improvements in per-unit efficiency. On the whole, however, the overall picture is the same: no matter what the gains in per-unit efficiency are—more widgets per work hour, more work per barrel of oil, more food per hectare of farmland—the overall effect of per-unit efficiency gains is growth in aggregate demand. That means that even as we get better at using less material and energy per unit of phone, laptop, or monitor that we produce (what is coming to be known as "lightweight-ing" in the industry; see Babbitt 2017; Staub 2017), overall demand for those materials and energy rebounds and increases—as do the conse-quences for discards from that overall growth in demand.

What the Jevons paradox suggests is that we must beware of what the ecological economist Blake Alcott calls our "inner engineer" (Alcott 2015). The inner engineer is characterized as striving to achieve more efficient use of resources as an end in itself. Pursuing that end gives rise to attempts to achieve overall reductions in resource demand through gains in per-unit efficiency. The Jevons paradox or the rebound effect tells us that such an environmental strategy is a false hope. Growth in aggregate demand will swamp gains in per-unit efficiency.

The Jevons paradox points to the need for political action that reorga-nizes extraction for, and manufacturing of, electronics based on the premise that the necessary resources are held in the commons (more generally, see Fennell 2011; Ostrom 1990; Ostrom et al. 1999). Degrowth offers neither a blueprint nor a plan to achieve that. It does, however, offer a sensibility—that embodied in the difference between efficiency versus sufficiency—that must inform whatever specific plans are proffered to achieve actual reduc-tions in total energy and material throughput at all points in the existence of electronics (e.g., legislating hard caps on mineral extraction; see Alcott

2015). The life work of the Nobel Prize–winning economist Elinor Ostrom documents successes in a variety of sectors (e.g., fisheries, forestry, water use) where the resources being extracted are done so in common and successfully over both long periods of time and substantial ranges of space. No template from those successes exists that might simply be transferred to the electronics sector. But if it exists, it is possible, and what is possible can always be organized otherwise, even if not easily so.

* * *

This book opened with a question—what is the right thing to do with electronic waste? I hope that I have demonstrated that there is more than one way to answer that question. Other answers may be waiting to be imagined; however, I would suggest that dealing with the waste that arises ubiquitously but unevenly at all points in the existence of electronics requires keeping three lessons in mind. First, hazard and waste are indeterminate categories. They are loaded with inherent ambiguities that cannot all be overcome with more and more precise definitions. Second, waste can also be noncoherent. Even in cases where precise definitions exist, what counts as waste, where, and according to whom can vary. And third, waste must be defamiliarized: the idea that solutions to problems of waste lie in the proper management of postconsumer waste alone needs to be reformatted as a peculiar, even an odd notion rather than a conventional one. If we are to handle waste from electronics with care (Puig de la Bellacasa 2011, 2012), then along with these lessons it might help to ask several more questions to organize our responses: How will we keep electronics going? *Should* we keep them going? If so, who and what must be considered; that is, whom and what do and should we care about and care for? No one yet knows all the answers to these questions. In that sense, I end on a note of invitation to readers to seek out answers to these sorts of questions in ways that no single author could possibly imagine alone.

References

Advocate General Jacobs. 1997. Opinion of Mr Advocate General Jacobs delivered on 24 October 1996. Criminal proceedings against Euro Tombesi and Adino Tombesi (C-304/94), Roberto Santella (C-330/94), Giovanni Muzi and others (C-342/94) and Anselmo Savini (C-224/95). European Court.

Ahmed, Syed Ishtiaque, Steven J. Jackson, and Md. Rashidujjaman Rifat. 2015. "Learning to Fix: Knowledge, Collaboration and Mobile Phone Repair in Dhaka, Bangladesh." In *ICTD* 15, May 15–18, 1–10. Singapore: ACM Press. doi:10.1145/2737856.2738018.

Akbar, Jay. 2015. "Africa's 'Electronic Graveyards' Where the West Dumps PCs, Laptops and More." *Daily Mail*, April 23. http://www.dailymail.co.uk/news/article-3049457/Where-computer-goes-die-Shocking-pictures-toxic-electronic-graveyards-Africa-West-dumps-old-PCs-laptops-microwaves-fridges-phones.html.

Akese, Grace Abena. 2014. "Price Realization for Electronic Waste (e-Waste) in Accra, Ghana." Master's thesis, Memorial University of Newfoundland. http://research.library.mun.ca/8112.

Akrich, Madeline. 1992. "The De-Scription of Technical Objects." In *Shaping Technology/Building Society*, ed. Wiebe E. Bijker and John Law, 205–224. Cambridge, MA: MIT Press.

Alcott, Blake. 2015. "Jevons' Paradox (Rebound Effect)." In *Degrowth: A Vocabularly for a New Era*, ed. Giacomo D'Alisa, Federico Demaria, and Giorgos Kallis, 121–124. New York: Routledge.

Amoyaw-Osei, Yaw, O. Agyekum, Esther Mueller, R. Fasko, and Mathias Schluep. 2011. "Ghana E-Waste Country Assessment." Secretariat of the Basel Convention. http://www.ewasteguide.info/Amoyaw-Osei_2011_GreenAd-Empa.

Apple. 2017. "Environmental Responsibility Report: 2017 Progress Report, Covering Fiscal Year 2016." Apple Computers Inc. https://images.apple.com/environment/pdf/Apple_Environmental_Responsibility_Report_2017.pdf.

Archambault, Éric, and Vincent Larivière. 2010. "The Limits of Bibliometrics for the Analysis of the Social Sciences and Humanities Literature." *International Social Science Council, World Social Sciences Report: Knowledge Divides.* Paris: UNESCO. http://www.ost.qc.ca/Portals/0/docs/articles/2010/Biblio_WorldSocScienceReport.pdf.

Arensman, Russ. 2000. "Ready for Recycling?" *Electronic Business* 26 (12): 108–115.

Babbitt, Callie W. 2017. "Overview of Slides [DRAFT]." http://www.cta.tech/cta/media/policyimages/rit-study.pdf.

Baldé, C. P., F. Wang, R. Kuehr, and J. Huisman. 2015. *The Global E-Waste Monitor 2014: Quantities, Flows and Resources.* Bonn: United Nations University.

Baldé, C.P., F. Wang, and Ruediger Kuehr. 2016. "Transboundary Movements of Used and Waste Electronic and Electrical Equipment." Bonn: United Nations University, StEP Initiative.

Bamako Secretariat. 1991. "Bamako Convention on the Ban of the Import into Africa and the Control of Transboundary Movement and Management of Hazardous Wastes within Africa." http://www.au.int/en/content/bamako-convention-ban-import-africa-and-control-transboundary-movement-and-management-hazard.

Barad, Karen. 2007. *Meeting the Universe Halfway: Quantum Physics and the Entanglement of Matter and Meaning.* Durham, NC: Duke University Press.

Barba-Gutiérrez, Y., B. Adenso-Díaz, and M. Hopp. 2008. "An Analysis of Some Environmental Consequences of European Electrical and Electronic Waste Regulation." *Resources, Conservation and Recycling* 52 (3): 481–495.

Basel Action Network. 2002. "Exporting Harm: The High-Tech Trashing of Asia." http://www.ban.org/main/library.html.

Basel Action Network. 2016a. "Best Buy Rolls Back Its Environmental Commitments." Press release, February 1. http://www.ban.org/news/2016/2/1/best-buy-rolls-back-its-environmental-commitments.

Basel Action Network. 2016b. "Child on_garbage_pic." Flickr page. https://www.flickr.com/photos/basel-action-network/9260624833.

Basel Action Network. 2016c. "Disconnect: Goodwill and Dell, Exporting the Public's E-Waste to Developing Countries." http://www.ban.org/trash-transparency.

Basel Action Network. 2016d. "E-Trash Transparency Project." http://www.ban.org/trash-transparency.

Basel Action Network. 2016e. "Staff." http://www.ban.org/staff.

Basel Secretariat. 2010. "Draft Technical Guidelines on Transboundary Movement of E-Waste, in Particular Regarding the Distinction between Waste and Non-Waste (Version 22 September 2010)." http://www.basel.int/Implementation/Ewaste/TechnicalGuidelines/DevelopmentofTGs/tabid/2377/Default.aspx.

Basel Secretariat. 2011a. "Basel Convention." http://www.basel.int/Portals/4/Basel%20Convention/docs/text/BaselConventionText-e.pdf.

Basel Secretariat. 2011b. "Draft Technical Guidelines on Transboundary Movement of E-Waste, in Particular Regarding the Distinction between Waste and Non-Waste (Version 21 February 2011)." http://archive.basel.int/techmatters/code/comments.php?guidId=78.

Basel Secretariat. 2012a. "Draft Technical Guidelines on Transboundary Movement of E-Waste, in Particular Regarding the Distinction between Waste and Non-Waste (Version 8 May 2012)." http://www.basel.int/Implementation/Ewaste/Technical Guidelines/DevelopmentofTGs/tabid/2377/Default.aspx.

Basel Secretariat. 2012b. "Draft Technical Guidelines on Transboundary Movement of E-Waste, in Particular Regarding the Distinction between Waste and Non-Waste (Version of 27 September 2012)." http://www.basel.int/Portals/4/download.aspx?d=UNEP-CHW-OEWG.8-INF-9-Rev.1.English.doc.

Basel Secretariat. 2014. "Draft Technical Guidelines on Transboundary Movements of Electronic and Electrical Waste and Used Electrical and Electronic Equipment, in Particular Regarding the Distinction between Waste and Non-Waste under the Basel Convention (Draft of 20 November 2014)." http://www.basel.int/Implementation/Ewaste/TechnicalGuidelines/DevelopmentofTGs/tabid/2377/Default.aspx.

Basel Secretariat. 2015a. "Decision BC-12/5: Technical Guidelines on Transboundary Movements of Electrical and Electronic Waste and Used Electrical and Electronic Equipment, in Particular Regarding the Distinction between Waste and Non-Waste under the Basel Convention." http://www.basel.int/TheConvention/Conference oftheParties/Meetings/COP12/tabid/4248/mctl/ViewDetails/EventModID/8051/EventID/542/xmid/13027/Default.aspx.

Basel Secretariat. 2015b. "Report of the Conference of the Parties to the Basel Convention on the Control of Transboundary Movements of Hazardous Wastes and Their Disposal on the Work of Its Twelfth Meeting." http://www.basel.int/The Convention/ConferenceoftheParties/Meetings/COP12/tabid/4248/mctl/View Details/EventModID/8051/EventID/542/xmid/13027/Default.aspx.

Basel Secretariat. 2015c. "Summary Record of the Physical Meeting of the Small Intersessional Working Group (SIWG) on E-Waste." http://www.basel.int/Portals/4/download.aspx?d=UNEP-CHW-TG-EWASTE-SAO-SwigSummary.English.docx.

Basel Secretariat. 2015d. "Twelfth Meeting of the Conference of the Parties to the Basel Convention." http://www.basel.int/TheConvention/ConferenceoftheParties/Meetings/COP12/tabid/4248/mctl/ViewDetails/EventModID/8051/EventID/542/xmid/13027/Default.aspx.

Basel Secretariat. 2017. "The Basel Convention BAN Amendment: Overview." http://www.basel.int/Implementation/LegalMatters/BanAmendment/Overview/tabid/1484/Default.aspx.

Bell, Genevieve, Mark Blythe, and Phoebe Sengers. 2005. "Making by Making Strange: Defamiliarization and the Design of Domestic Technologies." *ACM Transactions on Computer-Human Interaction* 12 (2): 149–173.

Bereisa, J. 1983. "Applications of Microcomputers in Automotive Electronics." *IEEE Transactions on Industrial Electronics* IE-30 (2): 87–96. doi:10.1109/TIE.1983.356715.

Blacksmith Institute. 2013. "The World's Worst 2013: The Top Ten Toxic Threats." New York: Blacksmith Institute. http://www.blacksmithinstitute.org/new-report -cites-the-world-s-worst-polluted-places.html.

Bloomberg. 2017. "These Women Are Paying the Price for Our Digital World." Bloomberg.com, June 15. https://www.bloomberg.com/news/photo-essays/2017-06 -15/these-women-are-paying-the-price-for-our-digital-world.

Bonham-Carter, G. F. 2005. "Introduction to the GSC MITE Point Sources Project," bulletin no. 584. http://geoscan.ess.nrcan.gc.ca/cgi-bin/starfinder/0?path=geoscan.fl &id=fastlink&pass=&search=R%3D221047&format=FLFULL.

Borg, Kevin L. 2007. *Auto Mechanics: Technology and Expertise in Twentieth-Century America.* Baltimore, MD: Johns Hopkins Univesity Press.

Bornman, Lutz. 2013. "Evaluations by Peer Review in Science." *Springer Science Reviews* 1:1–4.

Bowker, Geoffrey C., and Susan Leigh Star. 1999. *Sorting Things Out: Classification and Its Consequences.* Cambridge, MA: MIT Press.

Breivik, Knut, James Michael Armitage, Frank Wania, and Kevin C. Jones. 2014. "Tracking the Global Generation and Exports of E-Waste. Do Existing Estimates Add Up?" *Environmental Science & Technology* 48:8735–8743. doi:10.1021/es5021313.

Briggs, David F. 2015. "History of the Warren (Bisbee) Mining District." *Arizona Daily Independent*, June 4. https://arizonadailyindependent.com/2015/06/04/history-of -the-warren-bisbee-mining-district.

Brooks, Andrew. 2012. "Networks of Power and Corruption: The Trade of Japanese Used Cars to Mozambique." *Geographical Journal* 178 (1): 80–92. doi:10.1111/ j.1475-4959.2011.00410.x.

Broy, M., I. H. Kruger, A. Pretschner, and C. Salzmann. 2007. "Engineering Automotive Software." *Proceedings of the IEEE* 95 (2): 356–373. doi:10.1109/JPROC.2006.888386.

Bullard, Robert D. 1994. *Dumping in Dixie: Race, Class, and Environmental Quality.* 2nd ed. Boulder, CO: Westview Press.

Caley, David. 2007. "How to Think about Science? Interview with Simon Schaffer." *Ideas.* Canadian Broadcasting Corporation, January 2. http://www.cbc.ca/ideas/epi sodes/2009/01/02/how-to-think-about-science-part-1---24-listen.

Callén, Blanca, and Tomás Sánchez Criado. 2016. "Vulnerability Tests. Matters of 'Care for Matter' in E-Waste Practices." *TECNOSCIENZA: Italian Journal of Science & Technology Studies* 6 (2): 17–40.

Canadian Council of Ministers of the Environment. 2004. "Canada-wide Principles for Electronic Product Stewardship," ed. Canadian Council of Ministers of the Environment. Winnipeg, MB, June. http://www.ccme.ca/ourwork/waste.html.

Carr, Donald E. 1969. "Only the Giant Car-Eater Can Save Us: Government Subsidy Seems to Be in the Car-Disposal Picture." *New York Times*, May 4. http://query .nytimes.com/gst/abstract.html?res=9905E7D71E30EE3BBC4C53DFB3668382679 EDE.

Cass, Stephen. 2007. "How Much Does The Internet Weigh?" *Discover*, May 29. http://discovermagazine.com/2007/jun/how-much-does-the-internet-weigh.

Ceballos, Diana M., Wei Gong, and Elena Page. 2015. "A Pilot Assessment of Occupational Health Hazards in the US Electronic Scrap Recycling Industry." *Journal of Occupational and Environmental Hygiene* 12 (7): 482–488. doi:10.1080/15459624.201 5.1018516.

Chapman, Bruce. 2006. "Income Contingent Loans for Higher Education: International Reforms." In *Handbook of the Economics of Education*, ed. Eric Hanuscheck and Finis Welch, 2:1435–1497. Amsterdam: Elsevier.

Charette, Robert N. 2009. "This Car Runs on Code—IEEE Spectrum." http:// spectrum.ieee.org/green-tech/advanced-cars/this-car-runs-on-code.

Chaturvedi, Bharati. 2014. "A Waste of Wealth: How Indian Cities Are Ignoring the Recyclers but Asking for Recycling." *Environmental Justice* 7 (5): 138–141. doi:10.1089/ env.2014.0025.

Chen, Martha Alter. 2012. "The Informal Economy: Definitions, Theories and Policies." WIEGO Working Paper 1. Cambridge, MA: Women in Informal Employment: Globalizing and Organizing (WIEGO). http://wiego.org/publications/informal -economy-definitions-theories-and-policies.

Cheyne, Ilona, and Michael Purdue. 1995. "Fitting Definition to Purpose: The Search for a Satisfactory Definition of Waste." *Journal of Environmental Law* 7 (2): 149–168. doi:10.1093/jel/7.2.149.

Clapp, Jennifer. 1994. "Africa, NGOs, and the International Toxic Waste Trade." *Journal of Environment & Development* 3 (2): 17–46. doi:10.1177/107049659400300204.

Clark, Nigel, and Myra J. Hird. 2013. "Deep Shit." *O-Zone: A Journal of Object-Oriented Studies* 1:44–52.

Cook, Paul. 2017. *Secure E-Waste Export and Recycling Act* (H.R. 917). https://www .govtrack.us/congress/bills/115/hr917.

Corcoran, Patricia L., Charles J. Moore, and Kelly Jazvac. 2014. "An Anthropogenic Marker Horizon in the Future Rock Record." *GSA Today* (June): 4–8. doi:10.1130/GSAT-G198A.1.

CRS Holland. 2017. "CRS Holland—Cable Recovery Services." http://www.crsholland.com.

CWIT Project. 2016. "Project Overview—CWIT Project." http://www.cwitproject.eu/project-overview.

D'Alisa, Giacomo, Federico Demaria, and Giorgos Kallis, eds. 2014. *Degrowth: A Vocabulary for a New Era*. New York: Routledge.

Dannenberg, R. O., J. M. Maurice, and G. M. Potter. 1972. "Recovery of Precious Metal from Electronic Scrap," report no. 7683. Salt Lake City, UT: U.S. Department of the Interior, Bureau of Mines.

Dauvergne, Peter. 2016. *Environmentalism of the Rich*. Cambridge, MA: MIT Press.

DeLuca, Kevin Michael. 1999. *Image Politics: The New Rhetoric of Environmental Activism*. New York: Guilford Press.

Demos, T. J. 2013. *Return to the Postcolony: Specters of Colonialism in Contemporary Art*. Berlin: Sternberg Press.

Digital Methods Initiative. 2010. "Protocols Devised by the DMI." DigitalMethods.net, January 15. https://wiki.digitalmethods.net/Dmi/DmiProtocols.

Dillon, Lindsey. 2014. "Race, Waste, and Space: Brownfield Redevelopment and Environmental Justice at the Hunters Point Shipyard." *Antipode* 46 (5): 1205–1221. doi:10.1111/anti.12009.

DisplaySearch. 2014. "LCD TV Growth Improving, as Plasma and CRT TV Disappear." *Digital TV News*, April 16. http://www.digitaltvnews.net/?p=23994.

Douglas, Mary. 1966. *Purity and Danger: An Analysis of Concepts of Pollution and Taboo*. London: Routledge & Kegan Paul.

Duan, Huabo, Jiukun Hu, Quanyin Tan, Lili Liu, Yanjie Wang, and Jinhui Li. 2015. "Systematic Characterization of Generation and Management of E-Waste in China." *Environmental Science and Pollution Research International* (September): 1–15. doi:10.1007/s11356-015-5428-0.

Duan, Huabo, T. Reed Miller, Jeremy Gregory, and Randolph Kirchain. 2013. "Quantitative Characterization of Domestic and Transboundary Flows of Used Electronics: Analysis of Generation, Collection, and Export in the United States." UN Solve the E-waste Problem. http://r20.rs6.net/tn.jsp?f=001yyd1pMmZEhtbbcYnf6dD6zp5kZ9kD48clSblJr6cy8Q-RpMWiGvldxxxctOnRyILn3m55fQBy9iAG45rs2TDqvQ-SmBi6mRfJWcVeEoTWee65m85YB-xBQGD98mFbio6bG-6-LgzcMVKtajHZ8TfQU4JouAn6c31-TOFNcbXvW8=&c=jcEu2VKP_tmNcgSJia5pjx2DOpdB8AU0MDt8AHkUF

hux7G-aMJIm6g==&ch=RApSnwrzmD2oy2iaVdBFJVGvMeVxQyiC_ux6jtDVg6
SkOazw8IPLNA==.

Duan, Huabo, T. Reed Miller, Jeremy Gregory, and Randolph Kirchain. 2014. "Quantifying Export Flows of Used Electronics: Advanced Methods to Resolve Used Goods within Trade Data." *Environmental Science & Technology*, February. doi:10.1021/es404365z.

Economist. 2002. "A New Workshop of the World." *Economist*, October 10. http://www.economist.com/node/1382626.

Electronic Industry Citizenship Coalition. 2016. "Assessing and Reducing F-GHGs in the Electronics Supply Chain (March 2016)." EICC.org. http://www.eiccoalition.org/media/docs/publications/EICC_F-GHG_Report_2016.pdf.

EMPA. 2011. *Rapport technique de diagnostic national des mouvements transfrontières et de la gestion des DEEE—Benin.* Secretariat of the Basel Convention, October 31. http://www.basel.int/Portals/4/Basel%20Convention/docs/eWaste/E-waste_Africa_Project_Benin.pdf.

Environmental Investigation Agency. 2011. "System Failure: The UK's Harmful Trade in Electronic Waste." Environmental Investigation Agency. http://www.eia-international.org/files/news640-1.pdf.

E-Stewards and Basel Action Network. 2013. "Review Version: E-Stewards Standard for Responsible Recycling and Reuse of Electronic Equipment." http://e-stewards.org/learn-more/for-recyclers/access-the-standard/purchase-the-standard.

European Chemicals Agency. 2017. "Understanding REACH—ECHA." https://echa.europa.eu/regulations/reach/understanding-reach.

European Union. 1975. *Council Directive 75/442/EEC of 15 July 1975 on Waste.* Vol. Directive 75/442/EEC. http://eur-lex.europa.eu/legal-content/EN/TXT/HTML/?uri=CELEX:31975L0442&from=GA.

European Union. 1991. *Council Directive 91/156/EEC of 18 March 1991 Amending Directive 75/442/EEC on Waste.* Vol. Directive 91/156/EEC. http://eur-lex.europa.eu/legal-content/EN/TXT/?uri=CELEX%3A31991L0156.

European Union. 2006. *Regulation (EC) No 1013/2006 of the European Parliament and of the Council of 14 June 2006 on Shipments of Waste.* Vol. no 1013/2006. http://eur-lex.europa.eu/legal-content/EN/TXT/PDF/?uri=CELEX:02006R1013-20160101&rid=1.

European Union. 2007. "Revised Correspondents' Guidelines No 1. Shipments of Waste Electrical and Electronic Equipment (WEEE)."

Fairphone. 2013. "Tin and Tantalum Road Trip." Fairphone.com, November 8. https://www.fairphone.com/en/2013/11/08/tin-and-tantalum-road-trip.

Federal Highway Administration, U.S. Department of Transportation. 2017. "How the Highway Beautification Act Became a Law," June 27. https://www.fhwa.dot.gov/infrastructure/beauty.cfm.

Fennell, Lee. 2011. "Ostrom's Law: Property Rights in the Commons." *International Journal of the Commons* 5 (1). doi:10.18352/ijc.252.

Ferguson, James. 1999. *Expectations of Modernity: Myths and Meanings of Life on the Zambian Copperbelt.* Berkeley: University of California Press.

Feuz, Martin, Matthew Fuller, and Felix Stalder. 2011. "Personal Web Searching in the Age of Semantic Capitalism: Diagnosing the Mechanisms of Personalisation." *First Monday* 16 (2). http://firstmonday.org/ojs/index.php/fm/article/view/3344.

Finn, Bernard, and Daqing Yang, eds. 2009. *Communications under the Sea: The Evolving Cable Network and Its Implications.* Cambridge, MA: MIT Press.

Fitzpatrick, Colin, Elsa Olivetti, T. Reed Miller, Richard Roth, and Randolph Kirchain. 2015. "Conflict Minerals in the Computer Sector: Estimating Extent of Tin, Tantalum, Tungsten, and Gold Use in ICT Products." *Environmental Science & Technology* 49 (2): 974–981. doi:10.1021/es501193k.

Flink, James J. 1990. *The Automobile Age.* Cambridge, MA: MIT Press.

Fluck, Jurgen. 1994. "The Term Waste in EU Law." *European Environmental Law Review* 3:79–84.

Foote, Stephanie, and Elizabeth Mazzolini. 2012. *Histories of the Dustheap: Waste, Material Cultures, Social Justice.* Cambridge, MA: MIT Press.

Fosdick, Howard. 2012. "Computer Refurbishing: Environmentally Reducing the Digital Divide." *Bulletin of the American Society for Information Science and Technology* 38 (3): 58–62. doi:10.1002/bult.2012.1720380314.

Gabrys, Jennifer. 2011. *Digital Rubbish: A Natural History of Electronics.* Ann Arbor: University of Michigan Press.

Gallinatti, John, Deepa Gandhi, Eric Suchomel, Nancy Bice, Chuck Newell, and Poonam Kulkarni. 2012. "Managing a Large Dilute Plume Impacted by Matrix Diffusion: MEW Case Study." Paper presented at the Federal Remediation Technologies Roundtable, Washington, DC, June 20. https://frtr.gov/pdf/meetings/jun12/presentations/gallinatti-presentation.pdf.

Galvin, Ray. 2015. "The ICT/Electronics Question: Structural Change and the Rebound Effect." *Ecological Economics* 120 (December): 23–31. doi:10.1016/j.ecolecon.2015.08.020.

Georgescu-Roegen, Nicholas. 1970. "The Entropy Law and the Economic Problem." In *From Bioeconomics to Degrowth : Georgescu-Roegen's "New Economics" in Eight Essays,* 49–57. New York: Routledge.

Gephi Consortium. 2013. "Gephi, an Open Source Graph Visualization and Manipulation Software (version 0.8.2)." Mac OSX. http://gephi.org.

Gettleman, Jeffrey, and Marcus Bleasdale. 2013. "The Price of Precious." *National Geographic*, October. http://ngm.nationalgeographic.com/2013/10/conflict-minerals/gettleman-text.

Gill, Kaveri. 2009. *Of Poverty and Plastic: Scavenging and Scrap Trading Entrepreneurs in India's Urban Informal Economy*. New Delhi: OUP India.

Gille, Zsuzsa. 2007. *From the Cult of Waste to the Trash Heap of History: The Politics of Waste in Socialist and Postsocialist Hungary*. Bloomington: Indiana University Press.

Gille, Zsuzsa. 2013. "Is There an Emancipatory Ontology of Matter? A Response to Myra Hird." *Social Epistemology* 2 (4): 1–6.

Giridharadas, Anand. 2010. "Where a Cellphone Is Still Cutting Edge." *New York Times*, April 9. http://www.nytimes.com/2010/04/11/weekinreview/11giridharadas.html.

Global Witness. 2017. "Conflict Minerals." GlobalWitness.org. https://www.globalwitness.org/en/campaigns/conflict-minerals.

Google. 2011. *How Search by Image Works*. https://www.youtube.com/watch?v=keTZaJg0784.

Gorman, Alice. 2014. "The Anthropocene in the Solar System." *Journal of Contemporary Archaeology* 1 (1): 87–91.

GovTrack. 2009 "Electronic Waste Research and Development Act (2009—H.R. 1580)." GovTrack.us. https://www.govtrack.us/congress/bills/111/hr1580.

GovTrack. 2010. "H.R. 6252 (111th): Responsible Electronics Recycling Act." GovTrack.us. http://www.govtrack.us/congress/bill.xpd?bill=h111-6252.

Graham, Mark, and Stefano De Sabbata. 2016. "Information Geographies." http://geography.oii.ox.ac.uk/?page=home#.

Grant, Richard, and Martin Oteng-Ababio. 2012. "Mapping the Invisible and Real 'African' Economy: Urban E-Waste Circuitry." *Urban Geography* 33 (1): 1–21. doi:10.2747/0272-3638.33.1.1.

Gray-Cosgrove, Carmella, Max Liboiron, and Josh Lepawsky. 2015. "The Challenges of Temporality to Depollution & Remediation." *S.A.P.I.EN.S. Surveys and Perspectives Integrating Environment and Society* 8.1 (November). https://sapiens.revues.org/1740.

Greenpeace. 2017. "Clicking Clean." http://www.greenpeace.org/international/en/publications/Campaign-reports/Climate-Reports/clicking-clean-2017.

Gregory, Derek. 2009. "Geographical Imaginary." In *The Dictionary of Human Geography*. 5th ed., ed. Derek Gregory, Ron Johnston, Geraldine Pratt, Michael Watts, and Sarah Whatmore. Malden, MA: Wiley-Blackwell. http://qe2a-proxy.mun.ca/Login?url=http://search.credoreference.com/content/entry/bkhumgeo/geographical_imaginary/0.

Grimes, Seamus, and Yutao Sun. 2016. "China's Evolving Role in Apple's Global Value Chain." *Area Development and Policy* 1 (1): 94–112. doi:10.1080/23792949.2016.1149434.

Grossman, Elizabeth. 2006. *High Tech Trash: Digital Devices, Hidden Toxics, and Human Health*. Washington, DC: Island Press.

Grupo México and Southern Copper. 2014. "Integrated Operations." http://www.southernperu.com/ENG/intope/Pages/PGIntOperation.aspx.

Gu, Yifan, Yufeng Wu, Ming Xu, Huaidong Wang, and Tieyong Zuo. 2016. "The Stability and Profitability of the Informal WEEE Collector in Developing Countries: A Case Study of China." *Resources, Conservation and Recycling* 107 (February): 18–26. doi:10.1016/j.resconrec.2015.12.004.

Guardian. 2014. "Man Jailed for Illegally Exporting Electrical Waste to Africa." *Guardian*, June 20. http://www.theguardian.com/environment/2014/jun/20/man-jailed-illegal-exporting-electrical-waste-africa.

Gutberlet, Jutta. 2008. "Organized and Informal Recycling: Social Movements Contributing to Sustainability. Waste Management and the Environment IV." *WIT Transactions on Ecology and the Environment* 109:223–232. doi:10.2495/WM080241.

Haraway, Donna. 1991a. "Cyborgs at Large: Interview with Donna Haraway." In *Technoculture*, ed. Constance Penley and Andrew Ross, 1–21. Minneapolis: University of Minnesota Press.

Haraway, Donna. 1991b. *Simians, Cyborgs, and Women: The Reinvention of Nature*. New York: Routledge.

Haraway, Donna. 1997. *Modest_Witness@Second_Millenium.FemaleMan©_Meets_Onco-Mouse™*. New York: Routledge.

Haraway, Donna. 2008. *When Species Meet*. Minneapolis: University of Minnesota Press.

Haraway, Donna. 2010. "When Species Meet: Staying with the Trouble." *Environment and Planning D. Society & Space* 28 (1): 53–55.

Hird, Myra J. 2012. "Knowing Waste: Towards an Inhuman Epistemology." *Social Epistemology* 26 (3–4): 453–469. doi:10.1080/02691728.2012.727195.

Hird, Myra J. 2013. "Is Waste Indeterminacy Useful? A Response to Zsuzsa Gille." *Social Epistemology* 2 (6): 28–33.

Hirsch, Afua. 2013. "'This Is Not a Good Place to Live': Inside Ghana's Dump for Electronic Waste." *Guardian*, December 14. http://www.theguardian.com/world/2013/dec/14/ghana-dump-electronic-waste-not-good-place-live.

Höges, Clemens. 2009. "The Children of Sodom and Gomorrah: How Europe's Discarded Computers Are Poisoning Africa's Kids." *Spiegel*, December 4. http://www.spiegel.de/international/world/0,1518,665061,00.html.

Horner, Nathaniel C., Arman Shehabi, and Inês L. Azevedo. 2016. "Known Unknowns: Indirect Energy Effects of Information and Communication Technology." *Environmental Research Letters* 11 (10): 103001. doi:10.1088/1748-9326/11/10/103001.

Hu, Tung-Hui. 2015. *A Prehistory of the Cloud*. Cambridge, MA: MIT Press; https://mitpress.mit.edu/prehistory-cloud.

Huisman, Jaco, I. Botezatu, L. Herreras, M. Kiddane, J. Hintsa, V. Luda di Cortemiglia, P. Leroy, et al. 2015. "Countering WEEE Illegal Trade (CWIT) Summary Report, Market Assessment, Legal Analysis, Crime Analysis and Recommendations Roadmap." Lyon, France.

Hunter, Robert. 1971. *The Storming of the Mind*. Toronto: McClelland and Stewart.

IISD Reporting Services. 2017. "Summary of the Meetings of the Conferences of the Parties to the Basel, Rotterdam and Stockholm Conventions: 24 April–5 May 2017," *Earth Negotiations Bulletin*, May 8. http://enb.iisd.org/vol15/enb15252e.html.

Ingenthron, Robin. 2011. "Blog Reads Down: Externalizing Value." June. http://retroworks.blogspot.com/2011/06/blog-reads-down-externalizing-value.html.

Ingenthron, Robin. 2012. "PACE Yourself: Basel Logic of e-Waste Export." http://retroworks.blogspot.com/2012/03/pace-yourself-basel-convention-logic-of.html.

Ingenthron, Robin. 2017a. "Fair Trade Recycling (WR3A)." http://ingenthron.net/mission.html.

Ingenthron, Robin. 2017b. "How to Pay for Africa E-Waste Cleanup?" http://retroworks.blogspot.com/2017/02/how-to-pay-for-africa-e-waste-cleanup.html.

Ingenthron, Robin. 2017c. "How to Pay for Africa E-Waste Cleanup? Part 2." http://retroworks.blogspot.com/2017/02/how-to-pay-for-africa-e-waste-cleanup_5.html.

Ingenthron, Robin. 2017d. "How to Pay for Africa E-Waste Cleanup? Part 3." http://retroworks.blogspot.com/2017/02/how-to-pay-for-africa-e-waste-cleanup_9.html.

International Organization for Standards. 2013. *ISO 668: Series 1. Freight Containers—Classification, Dimensions and Ratings*. Geneva: International Organization for Standards.

INTERPOL. 2015. "Project Eden." http://www.interpol.int/Crime-areas/Environ mental-crime/Projects/Project-Eden.

Jackson, Steven J., Alex Pompe, and Gabriel Krieshok. 2012. "Repair Worlds: Maintenance, Repair, and ICT for Development in Rural Namibia." In *Proceedings of the ACM 2012 Conference on Computer Supported Cooperative Work*, 107–116. New York: ACM. http://dl.acm.org/citation.cfm?id=2145224.

Jameson, N. Jordan, Xin Song, and Michael Pecht. 2016. "Conflict Minerals in Electronic Systems: An Overview and Critique of Legal Initiatives." *Science and Engineering Ethics* 22 (5): 1375–1389.

Jensen, T. 2014. "Welfare Commonsense, Poverty Porn and Doxosophy." *Sociological Research Online* 19 (3). doi:10.5153/sro.3441.

Jones, Richard A. 2011. *Basel Action Network, Plaintiff, v. International Association of Electronics Recyclers, et al., Defendants*. U.S. District Court for the Western District of Washington.

JoshuaG . 2006. *Apple WWDC 2002-The Death of Mac OS 9*. https://www.youtube .com/watch?v=Cl7xQ8i3fc0.

Judge Dawson. 2014. *Regina v Benson*, sentencing remarks. Snaresbrook, UK.

Kahhat, Ramzy, and Eric Williams. 2012. "Materials Flow Analysis of E-Waste: Domestic Flows and Exports of Used Computers from the United States." *Resources, Conservation and Recycling* 67 (0): 67–74. doi:10.1016/j.resconrec.2012.07.008.

Kang, Hai-Yong, and Julie M. Schoenung. 2005. "Electronic Waste Recycling: A Review of U.S. Infrastructure and Technology Options." *Resources, Conservation and Recycling* 45 (4): 368–400.

Kaplan, Jeremy. 2014. "Welcome to Hell: Photographer Documents Africa's E-Waste Nightmare." Fox News, March 6. http://www.foxnews.com/tech/2014/03/06/ welcome-to-hell-photographer-documents-africas-e-waste-nightmare.html.

KCTS 9. 2016. "Behind the Scenes: How We Pulled This Off." KCTS 9—Public Television, May 9. http://kcts9.org/programs/circuit/behind-scenes-how-we-pulled.

Keeling, Arn. 2012. "Mineral Waste." In *Sage Encyclopedia of Consumption and Waste*, ed. Carl Zimring and William L. Rathje, 1:553–556. Thousand Oaks, CA: Sage.

Kellow, Aynsley. 1999. "Baptists and Bootleggers? The Basel Convention and Metals Recycling Trade." *Agenda (Durban, South Africa)* 6 (1): 29–38.

Kenahan, C. B., R. S. Kaplan, J. T. Dunham, and D. G. Linnehan. 1973. "Bureau of Mines Research Programs on Recycling and Disposal of Mineral-, Metal-, and Energy-Based Wastes." U.S. Bureau of Mines, Information Circular 8595.

Kim, Inah, Hyun J. Kim, Sin Y. Lim, and Jungok Kongyoo. 2012. "Leukemia and Non-Hodgkin Lymphoma in Semiconductor Industry Workers in Korea." *International Journal of Occupational and Environmental Health* 18 (2): 147–153. doi:10.11 79/1077352512Z.00000000019.

Kim, Inah, Myoung-Hee Kim, and Sinye Lim. 2015. "Reproductive Hazards Still Persist in the Microelectronics Industry: Increased Risk of Spontaneous Abortion and Menstrual Aberration among Female Workers in the Microelectronics Industry in South Korea." *PLoS One* 10 (5): 1–10. doi:10.1371/journal.pone.0123679.

Kim, Myoung-Hee, Hyunjoo Kim, and Domyung Paek. 2014. "The Health Impacts of Semiconductor Production: An Epidemiologic Review." *International Journal of Occupational and Environmental Health* 20 (2): 95–114. doi:10.1179/20493967 13Y.0000000050.

Kim, Yong H., and Gerald F. Davis. 2016. "Challenges for Global Supply Chain Sustainability: Evidence from Conflict Minerals Reports." *Academy of Management Journal* 59 (6): 1896–1916. doi:10.5465/amj.2015.0770.

Kleespies, E. K., J. P. Bennetts, and T. A. Henrie. 1970. "Gold Recovery from Scrap Electronic Solders by Fused-Salt Electrolysis." *Journal of Metals* 22 (1): 42–44.

Klein, Peter. 2009. "Ghana: Digital Dumping Ground." *Frontline*, PBS, June 23. http://www.pbs.org/frontlineworld/stories/ghana804/resources/ewaste.html.

Knapp, Freyja L. 2016. "The Birth of the Flexible Mine: Changing Geographies of Mining and the e-Waste Commodity Frontier." *Environment & Planning A* 48 (10): 1889–1909. doi:10.1177/0308518X16652398.

Koebler, Jason. 2017. "Apple Forces Recyclers to Shred All iPhones and MacBooks." *Motherboard*, April 20. https://motherboard.vice.com/en_us/article/apple-recycling -iphones-macbooks.

Koomey, Jonathan G. 2008. "Worldwide Electricity Used in Data Centers." *Environmental Research Letters* 3 (3). http://www.scopus.com/inward/record.url?eid =2-s2.0-54749110679&partnerID=40&md5=61d3b0f60289bedf3d2ac85f329e2532.

Koomey, Jonathan G. 2011. "Growth in Data Center Electricity Use 2005 to 2010." http://www.co.twosides.info/download/Koomey_Johnathon_G-_Growth_In_Data _Center_Electricity_Use_2005_to_2010_2011.pdf.

Koomey, Jonathan G., H. Scott Matthews, and Eric Williams. 2013. "Smart Everything: Will Intelligent Systems Reduce Resource Use?" *Annual Review of Environment and Resources* 38 (1): 311–343. doi:10.1146/annurev-environ-021512-110549.

Krulwich, Robert. 2011. "Let's Weigh the Internet (Or Maybe Let's Not)" (blog post). NPR, December 21. http://www.npr.org/blogs/krulwich/2011/12/21/144066248/lets -weigh-the-internet-or-maybe-lets-not.

Kummer Peiry, Katharina. 1992. "The International Regulation of Transboundary Traffic in Hazardous Wastes: The 1989 Basel Convention." *International and Comparative Law Quarterly* 41:530–562.

Kummer Peiry, Katharina. 1995. *International Management of Hazardous Wastes: The Basel Convention and Related Legal Rules*. Oxford: Oxford University Press.

La Rue, Frank, and United Nations. 2011. "Report of the Special Rapporteur on the Promotion and Protection of the Right to Freedom of Opinion and Expression." A/HRC/17/27. Geneva: United Nations. www2.ohchr.org/english/bodies/hrcouncil/docs/17session/A.HRC.17.27_en.pdf.

LaDou, J. 1984. "The Not-So-Clean Business of Making Chips." *Technology Review* 87 (4): 22–36.

LaDou, J. 1986. "Health Issues in the Microelectronics Industry." *Occupational Medicine-State of the Art Reviews* 1 (1): 1–11.

LaDou, J. 1991. "Semiconductor Industries." *Toxicology and Industrial Health* 7 (5–6): 59–62.

Latour, Bruno. 1987. *Science in Action*. Cambridge, MA: Harvard University Press.

Latour, Bruno. 1988. *The Pasteurization of France*. Cambridge, MA: Harvard University Press.

Latour, Bruno. 1993. *We Have Never Been Modern*. Cambridge, MA: Harvard University Press.

Latour, Bruno. 1999. *Pandora's Hope*. Cambridge, MA: Harvard.

Latour, Bruno. 2004. "Why Has Critique Run out of Steam? From Matters of Fact to Matters of Concern." *Critical Inquiry* 30 (2): 225–248.

Latour, Bruno. 2005. *Reassembling the Social: An Introduction to Actor Network Theory*. Oxford: Oxford University Press.

Latour, Bruno. 2010. "The Year in Climate Controversy." *Artforum* 49 (4): 228–229.

Latour, Bruno. 2012. "'What's the Story?' Organizing as a Mode of Existence." In *Agency without Actors? New Approaches to Collective Action*, ed. Jan-H. Passoth, Birgit Peuker, and Michael Schillmeier, 163–177. London: Routledge.

Latour, Bruno. 2015. "Telling Friends from Foes in the Time of the Anthropocene." In *The Anthropocene and the Global Environmental Crisis: Rethinking Modernity in a New Epoch*, ed. Christophe Bonneuil, Clive Hamilton, and François Gemenne. London: Routledge..

Latour, Bruno. n.d. "Introduction: How to Find One's Way in the Scientific Literature?"S02G01—Introduction. MediaLabSciencesPO. https://www.scribd.com/document/192542710/S02G01-introduction-S02-pdf.

Law, John. 2004a. *After Method: Mess in Social Science Research.* New York: Routledge.

Law, John. 2004b. "And If the Global Were Small and Noncoherent? Method, Complexity, and the Baroque." *Environment and Planning D. Society & Space* 22 (1): 13–26.

Law, John. 2010. "The Double Social Life of Method." http://www.heterogeneities .net/publications/Law2010DoubleSocialLifeofMethod5.pdf.

Law, John. 2012. "Collateral Realities." In *The Politics of Knowledge*, ed. Fernando Domínguez Rubio and Rubio Baert, 156–178. London: Routledge. 10.4324/9780203877746.

Lepawsky, Josh. 2012. "Legal Geographies of E-Waste Legislation in Canada and the US: Jurisdiction, Responsibility and the Taboo of Production." *Geoforum* 43 (6): 1194–1206.

Lepawsky, Josh. 2015a. "Are We Living in a Post-Basel World?" *Area* 47 (1): 7–15. doi:10.1111/area.12144.

Lepawsky, Josh. 2015b. "The Changing Geography of Global Trade in Electronic Discards: Time to Rethink the e-Waste Problem." *Geographical Journal* 181 (2): 147–159. doi:10.1111/geoj.12077.

Lepawsky, Josh. 2015c. "Trading on Distortion." *E-Scrap News*, December 22. http://resource-recycling.com/node/6803.

Lepawsky, Josh. 2016. "The Wake of the Anthropocene." *Techniques & Culture: Revue Semestrielle d'Anthropologie des Techniques*, October. http://tc.revues.org/7792.

Lepawsky, Josh. 2017. "Precision Is Not Accuracy." *Discard Studies*, January 12. https://discardstudies.com/2017/01/12/precision-is-not-accuracy.

Lepawsky, Josh, and Grace Abena Akese. 2015. "Sweeping Away Agbogbloshie. Again." *Discard Studies*, June 23. http://discardstudies.com/2015/06/23/sweeping -away-agbogbloshie-again.

Lepawsky, Josh, Erin Araujo, John-Michael Davis, and Ramzy Kahhat. 2017. "Best of Two Worlds? Towards Ethical Electronics Repair, Reuse, Repurposing and Recycling." *Geoforum* 81 (May): 87–99. doi:10.1016/j.geoforum.2017.02.007.

Lepawsky, Josh, and Mostaem Billah. 2011. "Making Chains That (Un)Make Things: Waste-Value Relations and the Bangladeshi Rubbish Electronics Industry." *Geografiska Annaler. Series B, Human Geography* 93 (2): 121–139. doi:10.1111/j.1468-0467 .2011.00365.x.

Lepawsky, Josh, and Creighton Connolly. 2016. "A Crack in the Facade? Situating Singapore in Global Flows of Electronic Waste." *Singapore Journal of Tropical Geography* 37 (2): 158–175. doi:10.1111/sjtg.12149.

Lepawsky, Josh, Joshua Goldstein, and Yvan Schulz. 2015. "Criminal Negligence?" *Discard Studies*, June 24. http://discardstudies.com/2015/06/24/criminal-negligence.

Lepawsky, Josh, and Chris McNabb. 2010. "Mapping International Flows of Electronic Waste." *Canadian Geographer/Le Géographe Canadien* 54 (2): 177–195. doi:10.1111/j.1541-0064.2009.00279.x.

Lewis, Geraint. 2014. "Where's the Proof in Science? There Is None." *The Conversation*, September 23. http://theconversation.com/wheres-the-proof-in-science-there-is-none-30570.

Li, J., J. Yang, and L. Liu. 2015. "Development Potential of E-Waste Recycling Industry in China." *Waste Management & Research* 33 (6): 533–542. doi:10.1177/07342 42X15584839.

Liboiron, Max. 2009. "Recycling As a Crisis of Meaning." *ETopia*. http://etopia. journals.yorku.ca/index.php/etopia/issue/view/2103/showToc.

Liboiron, Max. 2013a. "Modern Waste as Strategy." *Lo Squaderno: Explorations in Space and Society* 29:9–12.

Liboiron, Max. 2013b. "The Politics of Measurement: Per Capita Waste and Previous Sewage Contamination." *Discard Studies*, April 22. http://discardstudies.com/2013/04/22/the-politics-of-measurement-per-capita-waste-and-previous-sewage-contamination.

Liboiron, Max. 2014. "Against Awareness, for Scale: Garbage Is Infrastructure, Not Behavior." *Discard Studies*, January 23. http://discardstudies.com/2014/01/23/against-awareness-for-scale-garbage-is-infrastructure-not-behavior.

Liboiron, Max. 2015. "Redefining Pollution and Action: The Matter of Plastics." *Journal of Material Culture* (December): 1–24. doi:10.1177/1359183515622966.

Liboiron, Max. 2016. "Municipal versus Industrial Waste: Questioning the 3–97 Ratio." *Discard Studies*, March 2. http://discardstudies.com/2016/03/02/municipal-versus-industrial-waste-a-3-97-ratio-or-something-else-entirely.

Lippmann, Walter. 1925. *The Phantom Public*. New York: Harcourt, Brace.

Long, Keith R. 1995. "Production and Reserves of Cordilleran (Alaska to Chile) Porphyry Copper Deposits." *Arizona Geological Society Digest* 20:35–68.

Lord Justice Moses. 2008. *Environment Agency v Thorn International UK Ltd* (2008) EWHC 2595 (Admin). EWHC (Admin).

Lord Justice Pitchford. 2012. *Ezeemo & Ors v R.* (2012) EWCA Crim 2064. EWCA (Crim).

Lottermoser, Bernd. 2003. "Introduction to Mine Wastes." In *Mine Wastes*, 1–30. Springer. http://link.springer.com/chapter/10.1007/978-3-662-05133-7_1.

Lucsko, David N. 2016. *Junkyards, Gearheads, and Rust: Salvaging the Automotive Past.* Baltimore, MD: Johns Hopkins University Press.

Lüthje, Boy. 2007. "The Rise and Fall of 'Wintelism': Manufacturing Strategies and Transnational Production Networks of US Information Electronics Firms in the Pacific Rim." In *Competitveness of New Industries: Institutional Framework and Learning in Information Technology in Japan, the U.S. and Germany,* ed. Andreas Moerke, 180–209. New York: Routledge.

MacBride, Samantha. 2012. *Recycling Reconsidered: The Present Failure and Future Promise of Environmental Action in the United States.* Cambridge, MA: MIT Press.

Malmodin, Jens, Åsa Moberg, Dag Lundén, Göran Finnveden, and Nina Lövehagen. 2010. "Greenhouse Gas Emissions and Operational Electricity Use in the ICT and Entertainment & Media Sectors." *Journal of Industrial Ecology* 14 (5): 770–790. doi:10.1111/j.1530-9290.2010.00278.x.

Manhart, Andreas. 2010. "International Cooperation for Metal Recycling from Waste Electrical and Electronic Equipment." *Journal of Industrial Ecology* 15 (1): 13–30. doi:10.1111/j.1530-9290.2010.00307.x.

Manhart, Andreas, Osibanjo Osibanjo, Adeyinka Aderinto, and Siddharth Prakash. 2011. "Informal E-Waste Management in Lagos, Nigeria: Socio-Economic Impacts and Feasibility of International Recycling Co-Operations." Secretariat of the Basel Convention. http://www.basel.int/Portals/4/Basel%20Convention/docs/eWaste/E-waste_Africa_Project_Nigeria.pdf.

Mansfield, Roddy. 2012. "Eight Convicted of Toxic Waste Dump Scam." Sky News, December 5. http://news.sky.com/story/eight-convicted-of-toxic-waste-dump-scam-10461769.

Marres, Noortje. 2005. "Issues Spark a Public into Being." In *Making Things Public: Atmospheres of Democracy,* ed. Bruno Latour and Peter Weibel, 208–217. Cambridge, MA: MIT Press.

Marres, Noortje. 2012. "The Redistribution of Methods: On Intervention in Digital Social Research, Broadly Conceived." *Sociological Review* 60 (June): 139–165. doi:10.1111/j.1467-954X.2012.02121.x.

Marres, Noortje, and Carolin Gerlitz. 2015. "Interface Methods: Renegotiating Relations between Digital Social Research, STS and Sociology." *Sociological Review,* August. doi:10.1111/1467-954X.12314.

McCarthy, Tom. 2007. *Auto Mania: Cars, Consumers, and the Environment.* New Haven, CT: Yale University Press.

McElvaney, Kevin. 2014. "In Pictures: Ghana's e-Waste Magnet." Al Jazeera English, February 12. http://www.aljazeera.com/indepth/inpictures/2014/01/pictures-ghana-e-waste-mecca-2014130104740975223.html.

Meagher, Kate. 2013. "Unlocking the Informal Economy: A Literature Review on Linkages between Formal and Informal Economies in Developing Countries." WIEGO Working Paper 27. Cambridge, MA: Women in Informal Employment: Globalizing and Organizing. http://scholar.google.ca/scholar_url?hl=en&q=http://www.inclusivecities.org/wp-content/uploads/2013/06/Meagher_WIEGO_WP27.pdf&sa=X&scisig=AAGBfm2OJOCFN_EQFq8LIFgSg4cvzJInXQ&oi=scholaralrt.

Milmo, Cahal. 2009a. "How a Tagged Television Set Uncovered a Deadly Trade." *Independent*, February 18. http://www.independent.co.uk/news/world/africa/how-a-tagged-television-set-uncovered-a-deadly-trade-1624873.html.

Milmo, Cahal. 2009b. "Man Held after Tonnes of Illegal E-Waste Are Exported to Africa." *Independent*, November 7. http://www.independent.co.uk/news/uk/crime/man-held-after-tonnes-of-illegal-e-waste-are-exported-to-africa-1816570.html.

Minter, Adam. 2013a. *Junkyard Planet: Travels in the Billion-Dollar Trash Trade*. New York: Bloomsbury Publishing USA.

Minter, Adam. 2013b. "Scenes from a Junkyard Planet: The Car Plague." *Shanghai Scrap*, September 26. http://shanghaiscrap.com/2013/09/scenes-from-a-junkyard-planet-the-car-plague.

Minter, Adam. 2016. "The Burning Truth behind an E-Waste Dump in Africa." *Smithsonian Magazine*, January 13. http://www.smithsonianmag.com/science-nature/burning-truth-behind-e-waste-dump-africa-180957597.

Mol, Annemarie. 2002. *The Body Multiple: Ontology in Medical Practice*. Durham, NC: Duke University Press.

Möller, Jürgen. 2009. "Animal Feeding Stuff: Global Standard for the Determination of Acid Detergent Fibre (ADF) and Lignin." Foss.dk. http://www.allaboutfeed.net/Home/General/2009/3/Global-standard-for-determining-ADF-and-lignin-AAF011603W.

Mongabay. 2012. "High-Tech Hell: New Documentary Brings Africa's e-Waste Slum to Life." *Mongabay Environmental News*, April 30. http://news.mongabay.com/2012/04/high-tech-hell-new-documentary-brings-africas-e-waste-slum-to-life.

Montgomery, Mark A. 1995. "Reassessing the Waste Trade Crisis: What Do We Really Know?" *Journal of Environment & Development* 4 (1): 1–28. doi:10.1177/107049659500400102.

Myslicki, John. 2009. "Basel Convention 'From Cairo to Basel.'" February 24. http://www.umweltdaten.de/abfallwirtschaft/gav/Myslicki-From_Cairo_to_Basel.pdf.

Nalcor Energy. 2017. "Project Overview: Lower Churchill Project." https://muskrat falls.nalcorenergy.com/project-overview.

NASA Orbital Debris Program Office. 2017. *Orbital Debris Quarterly News* 21 (1). https://orbitaldebris.jsc.nasa.gov/Quarterly-News/pdfs/ODQNv21i1.pdf.

Nest, Michael Wallace. 2011. *Coltan*. Cambridge, UK: Polity.

Newman, N., C. Jones, E. Page, D. Ceballos, and A. Oza. 2015. "Investigation of Childhood Lead Poisoning from Parental Take-Home Exposure from an Electronic Scrap Recycling Facility—Ohio, 2012." *Morbidity and Mortality Weekly Report* 64 (27): 743–745.

Nordstrom, Carolyn. 2007. *Global Outlaws: Crime, Money, and Power in the Contemporary World*. Berkeley: University of California Press.

O'Brien, Mary H. 1993. "Being a Scientist Means Taking Sides." *Bioscience* 43 (10): 706–708. doi:10.2307/1312342.

Ogungbuyi, Olakitan, Innocent Chidi Nnorom, Oladele Osibanjo, and Mathias Schluep. 2012. "E-Waste Country Assessment Nigeria." Basel Convention Coordinating Centre, Nigeria / Swiss Federal Laboratories for Materials Science and Technology (EMPA), Switzerland. http://www.basel.int/Implementation/TechnicalAssistance/EWaste/EwasteAfricaProject/PublicationsReports/tabid/2553/Default.aspx#.

Ohkura, Makoto, Fan Luo, Eui-Guen Lee, Kunio Matsumura, Frank Niu, and Seok-Hyun Seong. 2012. "Fluorinated Compounds Emission Reduction Activity of WLICC (World LCD Industry Cooperation Committee)." In *SID Symposium Digest of Technical Papers* 43:838–841. Wiley Online Library. http://onlinelibrary.wiley.com/doi/10.1002/j.2168-0159.2012.tb05916.x/full.

Olivieri, Adam, Don Eisenberg, Martin Kurtovich, and Lori Pettegrew. 1985. "Ground-Water Contamination in Silicon Valley." *Journal of Water Resources Planning and* Management 111 (3). doi:10.1061/(ASCE)0733-9496(1985)111:3(346).

O'Neill, Kate. 2000. *Waste Trading among Rich Nations : Building a New Theory of Environmental Regulation.*Cambridge, MA: MIT Press.

onepercentfortheplanet.org. 2017. "How Does a Business Get Involved?" http://www.onepercentfortheplanet.org/what-we-do/our-approach/how-does-a-business-get-involved.

Ong, J. C. 2015. "Charity Appeals as 'Poverty Porn'? Production Ethics in Representing Suffering Children and Typhoon Haiyan Beneficiaries in the Philippines." In *Production Studies, The Sequel! Cultural Studies of Global Media Industries*, ed. Miranda Banks, Bridget Conor, and Vicki Mayer, 89–104. London: Routledge.

Oreskes, Naomi, and Erik M. Conway. 2010. *Merchants of Doubt: How a Handful of Scientists Obscured the Truth on Issues from Tobacco Smoke to Global Warming*. London: Bloomsbury.

Organisation for Economic Co-operation and Development. 2008. "Decision of the Council Concerning the Control of Transboundary Movements of Wastes Destined for Recovery Operations." http://acts.oecd.org/Instruments/ShowInstrumentView .aspx?InstrumentID=221&InstrumentPID=217&Lang=en&Book=False.

Ostrom, Elinor. 1990. *Governing the Commons: The Evolution of Institutions for Collective Action. Political Economy of Institutions and Decisions.* Cambridge: Cambridge University Press.

Ostrom, Elinor, Joanna Burger, Christopher B. Field, Richard B. Norgaard, and David Policansky. 1999. "Revisiting the Commons: Local Lessons, Global Challenges." *Science* 284 (5412): 278–282. doi:10.1126/science.284.5412.278.

Parikka, Jussi. 2015. *A Geology of Media.* Minneapolis: University of Minnesota Press.

Parks, Lisa. 2012. "When Satellites Fall: On the Trails of Cosmos 954 and USA193." In *Down to Earth: Satellite Technologies, Industries, and Cultures,* ed. Lisa Parks and James Schwoch, 221–237. New Brunswick, NJ: Rutgers University Press.

Parsons, Michael B., and Ray E. Cranston. 2006. "Influence of Lead Smelter Emissions on the Distribution of Metals in Marine Sediments from Chaleur Bay, Eastern Canada." *Geochemistry Exploration Environment Analysis* 6 (2–3): 259–275.

Partnership Africa Canada. 2017. "Conflict Minerals." http://www.pacweb.org/en/ conflict-minerals.

Pastides, H., E. J. Calabrese, D. W. Hosmer, Jr., and D. R. Harris, Jr. 1988. "Spontaneous Abortion and General Illness Symptoms among Semiconductor Manufacturers." *Journal of Occupational Medicine.* 30 (7): 543–551.

Pine, Kathleen H., and Max Liboiron. 2015. "The Politics of Measurement and Action." In *Proceedings of the 33rd Annual ACM Conference on Human Factors in Computing Systems,* 3147–3156. New York: ACM. doi:10.1145/2702123.2702298.

Propublica. 2015. "Nonprofit Explorer: Basel Action Network." https://projects .propublica.org/nonprofits/organizations/10918435.

Puca, Antonio, Marco Carrano, Gengyuan Liu, Dimitri Musella, Maddalena Ripa, Silvio Viglia, and Sergio Ulgiati. 2017. "Energy and EMergy Assessment of the Production and Operation of a Personal Computer." *Resources, Conservation and Recycling* 116 (January): 124–136. doi:10.1016/j.resconrec.2016.09.030.

Puckett, Jim, Leslie Byster, Sarah Westervelt, Richard Gutierrez, Sheia Davis, Asma Hussain, and Madhumitta Dutta. 2002. *Exporting Harm: The High-Tech Trashing of Asia.* Seattle: Basel Action Network, Silicon Valley Toxics Coalition. http://olo.ban .org/E-waste/technotrashfinalcomp.pdf.

Puckett, Jim, Sarah Westervelt, Richard Gutierrez, and Yuka Takamiya. 2005. "The Digital Dump: Exporting Re-Use and Abuse to Africa." Seattle: Basel Action Network. http://www.ban.org/BANreports/10-24-05/index.htm.

Puig de la Bellacasa, Maria. 2011. "Matters of Care in Technoscience: Assembling Neglected Things." *Social Studies of Science* 41 (1): 85–106. doi:10.1177/0306312710380301.

Puig de la Bellacasa, Maria. 2012. "'Nothing Comes without Its World': Thinking with Care." *Sociological Review* 60 (2): 197–216. doi:10.1111/j.1467-954X.2012.02070.x.

RAISE Hope for Congo. 2017. "Conflict Minerals." http://www.raisehopeforcongo.org/content/initiatives/conflict-minerals.

Rand, Lisa Ruth. 2014. "Gravity, the Sequel: Why the Real Story Would Be on the Ground." *Atlantic* (February 28). http://www.theatlantic.com/technology/archive/2014/02/-em-gravity-the-sequel-em-why-the-real-story-would-be-on-the-ground/284139.

Ratti, Carlo, Assaf Biderman, Dietmar Offenhuber, David Lee, Fábio Duarte, Brandon Nadres, David Perez, et al. 2016. "Monitour." http://senseable.mit.edu/monitour-app/#.

Reddy, Rajyashree N. 2013. "Revitalising Economies of Disassembly." *Economic and Political Weekly* 48 (13): 63.

Reddy, Rajyashree N. 2015. "Producing Abjection: E-Waste Improvement Schemes and Informal Recyclers of Bangalore." *Geoforum* 62 (June): 166–174. doi:10.1016/j.geoforum.2015.04.003.

Reddy, Rajyashree N. 2016. "Reimagining E-Waste Circuits: Calculation, Mobile Policies, and the Move to Urban Mining in Global South Cities." *Urban Geography* 37 (1): 57–76. doi:10.1080/02723638.2015.1046710.

Rekdal, Ole Bjørn. 2014. "Academic Urban Legends." *Social Studies of Science* 44 (June). doi:10.1177/0306312714535679.

repair.org. 2016. "The Association." *The Repair Association*.http://repair.org/association.

Rockwell, Geoffrey, and Stéfan Sinclair. 2016. *Hermeneutica: Computer-Assisted Interpretation in the Humanities*. Cambridge, MA: MIT Press.

Rogers, Richard. 2013. *Digital Methods*. Cambridge, MA: MIT Press.

Rogers, Richard, Natalia Sánchez-Querubín, and Aleksandra Kil. 2015. *Issue Mapping for an Ageing Europe*. Amsterdam: Amsterdam University Press. http://www.oapen.org/search?identifier=569806.

Rose, Gillian. 2001. *Visual Methodologies*. London: Sage.

Rosler, Martha. 2003. "In, around, and Afterthoughts." In *The Photography Reader*, ed. Liz Wells, 264–265. London: Routledge.

Rust, Susanne, and Matt Drange. 2014. "Cleanup of Silicon Valley Superfund Site Takes Environmental Toll." Emeryville, CA: Center for Investigative Reporting, March 17. https://www.revealnews.org/article/cleanup-of-silicon-valley-superfund -site-takes-environmental-toll-2.

Sander, Knut, Stephanie Schilling, Naoko Tojo, Chris van Rossem, Jan Vernon, and Carolyn George. 2007. "The Producer Responsibility Principle of the WEEE Direc- tive." ec.europa.eu/environment/waste/weee/pdf/final_rep_okopol.pdf.

Savard, Martine M., Graeme F. Bonham-Carter, and Catharine M. Banic. 2006. "A Geoscientific Perspective on Airborne Smelter Emissions of Metals in the Environ- ment: An Overview." *Geochemistry Exploration Environment Analysis* 6 (2–3): 99–109. doi:10.1144/1467-7873/05-095.

Sayer, Andrew. 2010. *Method in Social Science*, rev. 2nd ed. New York: Routledge.

SBS Australia and Giovana Vitola. 2011. "E-Waste Hell." *Giovana's Blog*. SBS, October 2. http://www.sbs.com.au/news/dateline/story/e-waste-hell.

Schartup, Amina T., Prentiss H. Balcom, Anne L. Soerensen, Kathleen J. Gosnell, Ryan S. D. Calder, Robert P. Mason, and Elsie M. Sunderland. 2015. "Freshwater Discharges Drive High Levels of Methylmercury in Arctic Marine Biota." *Proceedings of the National Academy of Sciences of the United States of America* 112 (38): 11789– 11794. doi:10.1073/pnas.1505541112.

Schluep, Mathias, Christian Hagelüken, Ruediger Kuehr, Federico Magalini, Claudia Maurer, Christina Meskers, Esther Mueller, and Feng Wang. 2009. "Recycling: From E-Waste to Resources." United Nations, July. https://www.scribd.com/document/ 117563731/E-Waste-Publication-Screen-FINALVERSION-Sml-Parte1.

Schluep, Mathias, Andreas Manhart, Osibanjo Osibanjo, David Rochat, Nancy Isarin, and Esther Mueller. 2011. *Where Are WEEE in Africa? Findings from the Basel Conven- tion E-Waste Africa Programme*. Secretariat of the Basel Convention. http://www.basel .int/Portals/4/download.aspx?d=UNEP-CHW-EWASTE-PUB-WeeAfricaReport .English.pdf.

Schmelzer, Matthias. 2016. *The Hegemony of Growth: The OECD and the Making of the Economic Growth Paradigm*. Cambridge: Cambridge University Press.

Schmidt, C. W. 1999. "Trading Trash: Why the US Won't Sign on to the Basel Con- vention." *Environmental Health Perspectives* 107 (8): A410–A412.

Schmitt, Robert J. 1990. "Automobile Shredder Residue: The Problem and Potential Solutions." CMP Report 90-1. Pittsburgh: Carnegie Mellon Research Institute, Center for Materials Production. http://infohouse.p2ric.org/ref/29/28953.pdf.

Schwartz, Joan M., and James R. Ryan. 2003. *Picturing Place: Photography and the Geographical Imagination*. London: I.B. Tauris.

Schwarzer, Stefan, Andréa De Bono, Gregory Giuliani, Stéfane Kluser, and Pascal Peduzzi. 2005. *E-Waste, the Hidden Side of IT Equipment's Manufacturing and Use*, Environment Alert Bulletin 5. United Nations Environment Programme. https://archive-ouverte.unige.ch/unige:23132.

Seaman, William. 2007. "Artificial Habitats and the Restoration of Degraded Marine Ecosystems and Fisheries." *Hydrobiologia* 580 (1): 143–155. doi:10.1007/s10750-006-0457-9.

Secretariat of the Basel Convention. 2011. "Basel Convention:Overview." http://www.basel.int/Implementation/Ewaste/EwasteinAfrica/Overview/tabid/2546/Default.aspx.

Seitz, Russell. 2007. "Weighing the Web." *Adamant* (blog), April 9. http://adamant.typepad.com/seitz/2007/04/weighing_the_we.html.

Sekula, Allan. 1978. "Dismantling Modernism, Reinventing Documentary (Notes on the Politics of Representation)." *Massachusetts Review* 19 (4): 859–883.

Self, Will. 2013. "Will Self Reads 'On Exactitude in Science' by Jorge Luis Borges." The Guardian Books podcast, January 4. Presented by Lisa Allardice and produced by Tim Maby. http://www.theguardian.com/books/audio/2013/jan/04/will-self-jorge-luis-borges.

SERI. 2014. "R2:2013: The Responsible Recycling ('R2') Standard for Electronics Recyclers." https://sustainableelectronics.org/r2-standard/r2-document-library.

Shapin, Steve, and Simon Schaffer. 1987. *Leviathan and the Air-Pump: Hobbes, Boyle, and the Experimental Life*. Princeton, NJ: Princeton University Press.

Simmons, Jerry. 2009. "Documerica: Snapshots of Crisis and Cure in the 1970s." *Prologue Magazine*. 41 (1). http://www.archives.gov/publications/prologue/2009/spring/documerica.html.

Simpson, Cam. 2017. "American Chipmakers Had a Toxic Problem. Then They Outsourced It." Bloomberg.com, June 15. https://www.bloomberg.com/news/features/2017-06-15/american-chipmakers-had-a-toxic-problem-so-they-outsourced-it.

Sinclair, Stéfan, and Geoffrey Rockwell. 2016. "Voyant Tools Help: About." http://voyant-tools.org/docs/#!/guide/about.

Sismondo, Sergio. 2017. "Post-Truth?" *Social Studies of Science* 47 (1): 3–6. doi:10.1177/0306312717692076.

Smith, Ted, Luke W. Cole, and Carl Wilmsen. 2003. *Ted Smith: Pioneer Activist for Environmental Justice in Silicon Valley, 1967–2000: Oral History Transcript*. Berkeley: University of California Libraries; http://archive.org/details/pioneeractivistsil00smitrich.

Smith, Ted, David Allan Sonnenfeld, and David N. Pellow. 2006. *Challenging the Chip: Labor Rights and Environmental Justice in the Global Electronics Industry*. Philadelphia: Temple University Press.

Souppouris, Aaron. 2016. "The Big Picture: The Human Cost of Electronic Waste." *Engadget*, February 18. http://www.engadget.com/2016/02/18/the-big-picture-e -waste.

Starosielski, Nicole. 2015. *The Undersea Network*. Durham, NC: Duke University Press.

Statistics Canada. 2012. "Human Activity and the Environment: Sections—Waste Management in Canada." 16–201–X. Ottawa: Statistics Canada. http://www.statcan. gc.ca/pub/16-201-x/2012000/parts-parties-eng.htm.

Staub, Colin. 2017. "How Lightweighting Has Shaken up the Electronics Stream." *E-Scrap*, May 4. https://resource-recycling.com/e-scrap/2017/05/04/lightweighting -shaken-electronics-stream.

Strasser, Susan. 1999. *Waste and Want: A Social History of Trash*. New York: Metropolitan Books.

Strohm, Laura A. 1993. "The Environmental Politics of the International Waste Trade." *Journal of Environment & Development* 2 (2): 129–153. doi:10.1177/ 107049659300200209.

Strother, James M., Henry O. Williams, and Mathias Schluep. 2012. "Used and End-of-Life Electrical and Electronic Equipment Imported into Liberia." United Nations Environmental Programme. http://www.basel.int/Portals/4/Basel%20Convention/ docs/eWaste/E-waste_Africa_Project_NationalAssessment_Liberia.pdf.

Tan, Xiao-Xin, Xiao-Jun Luo, Xiao-Bo Zheng, Zong-Rui Li, Run-Xia Sun, and Bi-Xian Mai. 2016. "Distribution of Organophosphorus Flame Retardants in Sediments from the Pearl River Delta in South China." *Science of the Total Environment* 544 (February): 77–84. doi:10.1016/j.scitotenv.2015.11.089.

Tao, Xiao-Qing, Dong-Sheng Shen, Jia-Li Shentu, Yu-Yang Long, Yi-Jian Feng, and Chen-Chao Shen. 2014. "Bioaccessibility and Health Risk of Heavy Metals in Ash from the Incineration of Different E-Waste Residues." *Environmental Science and Pollution Research International* 22 (5): 3558–3569. doi:10.1007/s11356-014-3562-8.

TeleGeography. 2017. "Submarine Cable FAQs." http://www2.telegeography.com/ submarine-cable-faqs-frequently-asked-questions.

Thomas, Courtney I. P. 2014. *Food We Trust: The Politics of Purity in American Food Regulation*. Lincoln: University of Nebraska Press.

Thorpe, B. 1992. "Public-Participation for Cleaner Production." *Nature & Resources* 28 (4): 38–44.

Trevizo, Perla. 2014. "Livelihoods Washed Away by Toxic Spill in Sonora." *Arizona Daily Star*, October 4. http://tucson.com/news/local/livelihoods-washed-away-by -toxic-spill-in-sonora/article_5b8007ef-82f1-5db1-901f-c4fba8cc1b06.html.

Turner, Chris. 2007. *The Geography of Hope: A Tour of the World We Need*. Toronto: Random House.

UBC News. 2010. "UBC Graduate School of Journalism Wins Emmy Award for Outstanding Investigative Journalism." *UBC News*, September 28. http://news .ubc.ca/2010/09/27/ubc-graduate-school-of-journalism-wins-emmy-award-for -outstanding-investigative-journalism.

UK Environment Agency. 2012. "Hazardous Waste Exporters Must Pay Six Figure Sum." http://www.repic.co.uk/files/Operation_Boron_-_Hazardous_waste_export- ers_sentencing_06_12_12_doc_doc.pdf.

UK Environment Agency. 2014. "Waste Dealer Jailed for 16 Months after Dangerous Shipments Stopped at Port." Press release, UK government, June 20. https://www .gov.uk/government/news/waste-dealer-jailed-for-16-months-after-dangerous -shipments-stopped-at-port.

United Kingdom. 2007. *The Transfrontier Shipment of Waste Regulations 2007*. Vol. 2007 No. 1711. www.legislation.gov.uk/uksi/2007/1711/pdfs/uksi_20071711_en .pdf.

United Nations Environmental Programme. 2009. "Submarine Cables and the Oceans: Connecting the World." https://www.unep-wcmc.org/resources-and-data/ submarine-cables-and-the-oceans--connecting-the-world.

United Nations Environmental Programme. 2015. "Waste Crime—Waste Risks | Gaps in Meeting the Global Waste Challenge." Nairobi: United Nations Environmental Programme. http://www.grida.no/_cms/OpenFile.aspx?s=1&id=1778.

United Nations StEP. 2014. "StEP E-Waste World Map." June 25. http://www.step -initiative.org/step-e-waste-world-map.html.

United Nations StEP. 2017a. "StEP E-Waste WorldMap: Overview Mexico." http:// www.step-initiative.org/Overview_Mexico.html.

United Nations StEP. 2017b. "StEP E-Waste World Map: Overview USA." http://www. step-initiative.org/Overview_USA.html.

U.S. Census Bureau. 2012. "Industry Snapshot: Automotive Mechanical and Electri- cal Repair and Maintenance (NAICS 81111)." http://thedataweb.rm.census.gov/The DataWeb_HotReport2/econsnapshot/2012/snapshot.hrml?NAICS=81111.

U.S. Congress. 2009. *Electronic Waste: Investing in Research and Innovation to Reuse, Reduce, and Recycle*. Washington, DC. http://www.gpo.gov/fdsys/pkg/CHRG -111hhrg47543/html/CHRG-111hhrg47543.htm.

U.S. Congress. 2010. *Dodd-Frank Wall Street Reform and Consumer Protection Act*. http://www.gpo.gov/fdsys/pkg/PLAW-111publ203/pdf/PLAW-111publ203.pdf.

U.S. Environmental Protection Agency. 1989. "Groundwater Contamination Clean-ups at South Bay Superfund Sites." San Francisco: U.S. Environmental Protection Agency. https://nepis.epa.gov/Exe/ZyPURL.cgi?Dockey=9100976C.txt.

U.S. Environmental Protection Agency. 2011. "Toxicological Review of Trichloroeth-ylene." Washington, DC. http://cfpub.epa.gov/ncea/iris/iris_documents/documents/toxreviews/0199tr/0199tr.pdf.

U.S. Environmental Protection Agency. 2012a. "A Citizen's Guide to Activated Carbon Treatment." Washington, DC. https://www.epa.gov/remedytech/citizens-guide -activated-carbon-treatment.

U.S. Environmental Protection Agency. 2012b. "A Citizen's Guide to Air Stripping." Washington, DC. https://www.epa.gov/remedytech/citizens-guide-air-stripping.

U.S. Environmental Protection Agency. 2012c. "A Citizen's Guide to Pump and Treat." Washington, DC. https://www.epa.gov/remedytech/citizens-guide-pump -and-treat.

U.S. Environmental Protection Agency. 2016a. "Center for Corporate Climate Lead-ership Sector Spotlight: Electronics." Overviews and Factsheets. Washington, DC. https://www.epa.gov/climateleadership/center-corporate-climate-leadership-sector -spotlight-electronics.

U.S. Environmental Protection Agency. 2016b. "F-GHG Emissions Reduction Efforts: 2016 Flat Panel Display Supplier Profiles." Overviews and Factsheets. Washington, DC, September. https://www.epa.gov/climateleadership/f-ghg-emissions-reduction -efforts-2016-flat-panel-display-supplier-profiles-0.

U.S. Environmental Protection Agency. 2017a. "Site Overviews, Middlefield-Ellis-Whisman (MEW) Study Area, U.S. EPA, Pacific Southwest, Superfund." Overviews and Factsheets. Washington, DC, February 21. https://yosemite.epa.gov/r9/sfund/r9sfdocw.nsf/ce6c60ee7382a473882571af007af70d/e4b75798264cff7988257007005e946e!OpenDocument.

U.S. Environmental Protection Agency. 2017b. "TRI Around the World." Overviews and Factsheets. March 24. https://www.epa.gov/toxics-release-inventory-tri-program/tri-around-world.

U.S. Environmental Protection Agency. 2017c. "TRI Pollution Prevention (P2) Search." Data and Tools. https://www3.epa.gov/enviro/facts/tri/p2.html.

U.S. Food and Drug Administration. 2002. "Science and the Regulation of Biological Products: From a Rich History to a Challenging Future." Silver Springs, MD: U.S. Food and Drug Administration. https://www.fda.gov/downloads/AboutFDA/What WeDo/History/ProductRegulation/100YearsofBiologicsRegulation/UCM070313.pdf.

U.S. Geological Survey. 1991. "Lavender Pit (MRDS #10046238) CU." February 1. https://mrdata.usgs.gov/mrds/show-mrds.php?dep_id=10046238.

U.S. Geological Survey and Thomas G. Goonan. 2005. "Flows of Selected Materials Associated with World Copper Smelting." Reston, VA: U.S. Geological Survey. https://pubs.usgs.gov/of/2004/1395.

U.S. Geological Survey, T. D. Kelly, and G. R. Matos. 2005. "Copper End-Use Statistics." Reston, VA: U.S. Geological Survey. https://minerals.usgs.gov/minerals/pubs/historical-statistics/copper-use.xls.

Valverde, Mariana. 2009. "Jurisdiction and Scale: Legal 'Technicalities' as Resources for Theory." *Social & Legal Studies* 18 (2): 139–157. doi:10.1177/0964663909103622.

van der Stap, Tim, Joop W. P. Coolen, and Han J. Lindeboom. 2016. "Marine Fouling Assemblages on Offshore Gas Platforms in the Southern North Sea: Effects of Depth and Distance from Shore on Biodiversity." *PLoS One* 11 (1): e0146324. doi:10.1371/journal.pone.0146324.

Venturini, Tommaso. 2010. "Diving in Magma: How to Explore Controversies with Actor-Network Theory." *Public Understanding of Science* 19 (3): 258–273. doi:10.1177/0963662509102694.

Venturini, Tommaso. 2012. "Building on Faults: How to Represent Controversies with Digital Methods." *Public Understanding of Science* 21 (7): 796–812. doi:10.1177/0963662510387558.

Victor, Peter A. 2008. *Managing without Growth: Slower by Design, Not Disaster.* Cheltenham: Edward Elgar.

Vogel, Christoph, and Timothy Raeymaekers. 2016. "Terr(It)or(Ies) of Peace? The Congolese Mining Frontier and the Fight against 'Conflict Minerals,'" *Antipode* 48 (4): 1102–1121. doi:10.1111/anti.12236.

Wang, F., J. Huisman, C. E. M. Meskers, M. Schluep, A. Stevels, and C. Hagelüken. 2012. "The Best-of-2-Worlds Philosophy: Developing Local Dismantling and Global Infrastructure Network for Sustainable e-Waste Treatment in Emerging Economies." *Waste Management* 32 (11): 2134–2146.

Wang, Yalin, Jinxing Hu, Wei Lin, Ning Wang, Cheng Li, Peng Luo, Muhammad Zaffar Hashmi, et al. 2016. "Health Risk Assessment of Migrant Workers' Exposure to Polychlorinated Biphenyls in Air and Dust in an e-Waste Recycling Area in China: Indication for a New Wealth Gap in Environmental Rights." *Environment International* 87 (February): 33–41. doi:10.1016/j.envint.2015.11.009.

Washington Materials Management and Financing Authority. 2015. "E-Cycle Washington Standard Plan 2015 Annual Report: Workbook Format." Washington, DC: Washington Materials Management & Financing Authority. http://www.ecy.wa.gov/programs/swfa/eproductrecycle/docs/2015WMMFAAnnualReport.pdf.

Wernink, Tessa, and Carina Strahl. 2015. "Fairphone: Sustainability from the Inside-Out and Outside-In." In *Sustainable Value Chain Management*, ed. Michael D'heur, 123–139. Heidelberg: Springer International Publishing. http://link.springer.com/chapter/10.1007/978-3-319-12142-0_3.

Whitford, David. 2017. "Meet the $21 Million Company That Thinks a New IPhone Is a Total Waste of Money." Inc.com, April. https://www.inc.com/magazine/201704/david-whitford/ifixit-repair-men.html.

Wiens, Kyle. 2015. "We Won Exemptions for Repairing Tractors, Cars, and Tablets." *IFixit*, October 7. https://ifixit.org/blog/7475/repair-coalition-wins-exemptions.

Williams, Eric. 2004. "Energy Intensity of Computer Manufacturing: Hybrid Assessment Combining Process and Economic Input-Output Methods." *Environmental Science & Technology* 38 (22): 6166–6174.

Williams, Eric, R. U. Ayres, and M. Heller. 2002. "The 1.7 Kilogram Microchip: Energy and Material Use in the Production of Semiconductor Devices." *Environmental Science & Technology* 36 (24): 5504–5510.

Winter, Caroline. 2013. "Why It's Hard to Make an Ethically Sourced Smartphone." *BusinessWeek*, September 20. http://www.businessweek.com/articles/2013-09-20/why-its-hard-to-make-an-ethically-sourced-smartphone.

Woolgar, Steve. 2011. *Where Did All the Provocation Go? Reflections on the Fate of Laboratory Life (1979)*. http://www.youtube.com/watch?v=gPDNptLkiyk&feature=youtube_gdata_player.

World Bank. 2017. "Container Port Traffic (TEU: 20 Foot Equivalent Units): Ghana." http://data.worldbank.org/indicator/IS.SHP.GOOD.TU?locations=GH.

WorldLoop. 2013. "About WorldLoop." http://worldloop.org/about-worldloop.

Wynne, Brian. 1987. *Risk Management and Hazardous Waste: Implementation and the Dialectics of Credibility*. London: Springer.

Wynne, Brian. 1989. "The Toxic Waste Trade: International Regulatory Issues and Options." *Third World Quarterly* 11 (3): 120–146.

Xiezhi, Yu, Zennegg Markus, Engwall Magnus, Rotander Anna, Larrson Maria, Wong Ming Hung, and Weber Roland. 2008. "E-Waste Recycling Heavily Contaminates a Chinese City with Chlorinated, Brominated and Mixed Halogenated Dioxins." *Organohalogen Compounds* 70:813–816.

Xu, P., X. Lou, G. Ding, H. Shen, L. Wu, Z. Chen, J. Han, and X. Wang. 2015. "Effects of PCBs and PBDEs on Thyroid Hormone, Lymphocyte Proliferation, Hematology and Kidney Injury Markers in Residents of an e-Waste Dismantling Area in Zhejiang, China." *Science of the Total Environment* 536:215–222. doi:10.1016/j.scitotenv.2015.07.025.

Xu, Xijin, Xiang Zeng, H. Marike Boezen, and Xia Huo. 2015. "E-Waste Environmental Contamination and Harm to Public Health in China." *Frontiers of Medicine* (March): 1–9. doi:10.1007/s11684-015-0391-1.

Zhang, Q., J. Ye, J. Chen, H. Xu, C. Wang, and M. Zhao. 2014. "Risk Assessment of Polychlorinated Biphenyls and Heavy Metals in Soils of an Abandoned E-Waste Site in China." *Environmental Pollution* 185:258–265. doi:10.1016/j.envpol.2013.11.003.

Zimring, Carl A. 2005. *Cash for Your Trash: Scrap Recycling in America*. New Brunswick, NJ: Rutgers University Press.

Zimring, Carl A. 2011. "The Complex Environmental Legacy of the Automobile Shredder." *Technology and Culture* 52 (3): 523–547. doi:10.1353/tech.2011.0117.

Zimring, Carl A. 2016. *Clean and White: A History of Environmental Racism in the United States*. New York: NYU Press.

Index